T0332457

PLANT
FAMILIES

Simon Maughan has worked for fifteen years as an editor and publisher for the Royal Horticultural Society in the United Kingdom. Ross Bayton has a PhD in plant taxonomy and is a former editor of *BBC Gardeners' World*.

The University of Chicago Press, Chicago 60637
© 2017 Quarto Publishing plc

Conceived, Designed and Produced by The Bright Press,
an imprint of The Quarto Group
1 Triptych Place, London,
SE1 9SH, United Kingdom
T (0) 20 7700 6700 **F** (0)20 7700 8066
www.quarto.com

Published 2017
Printed in China

26 25 24 23 6

ISBN-13: 978-0-226-52308-8 (cloth)

ISBN-13: 978-0-226-53667-5 (e-book)

DOI: 10.7208/chicago/9780226536675.001.0001

Library of Congress Cataloging-in-Publication Data
Names: Bayton, Ross, author. | Maughan, Simon, author.
Title: Plant families : a guide for gardeners and botanists / Ross Bayton and Simon Maughan.
Description: Chicago : The University of Chicago Press, 2017.
| Includes bibliographical references and index.
Identifiers: LCCN 2017025968 | ISBN 9780226523088 (cloth : alk. paper)
| ISBN 9780226536675 (e-book)
Subjects: LCSH: Plants—Classification—Handbooks, manuals, etc.
| Botany—Handbooks, manuals, etc.
Classification: LCC QK95 .B378 2017 | DDC 580.1/2—dc23 LC record available at
https://lccn.loc.gov/2017025968

∞ This paper meets the requirements of ANSI/NISO Z39.48-1992
(Permanence of Paper).

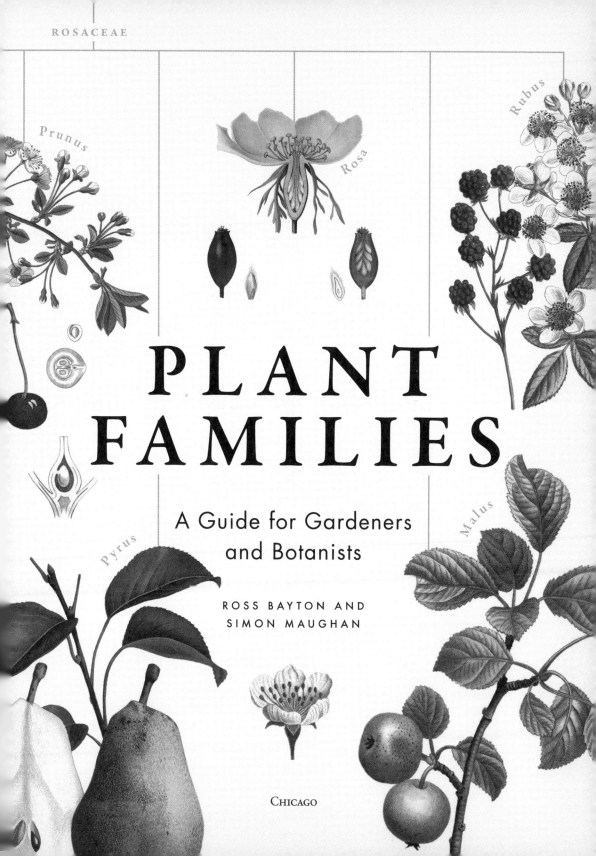

ROSACEAE

Prunus

Rosa

Rubus

PLANT FAMILIES

A Guide for Gardeners and Botanists

Pyrus

Malus

ROSS BAYTON AND
SIMON MAUGHAN

CHICAGO

Contents

Cycas balansae
in the cycad family
(*Cycadaceae*—see
pages 48–49).

CHAPTER 1

GYMNOSPERMS

CHAPTER 2

MONOCOTS AND EARLY ANGIOSPERMS

Magnolia liliiflora
in the magnolia family
(*Magnoliaceae*—see
pages 70–71).

EUDICOTS

Rosa rugosa in the rose family (*Rosaceae*—see pages 130–133).

Preface

While most of us think of plants as belonging to one big happy family, the fact is they don't. There are hundreds of different plant families, which botanists have cleverly grouped together, using what they know of family histories and genealogy, to bring some sense and order to more than a quarter of a million different plant species.

Plant families are all around us. Whatever the time of year, go for a walk and look for wild or garden plants. You'll be surprised at how many plant families are represented within a small radius of your home. Even in your own yard there will be a fantastic genealogy of plants, thanks largely to the efforts of plant collectors and horticulturists who, over the centuries, brought plants into cultivation from the four corners of the world.

When it comes to being a good gardener, making connections is what it is all about. If you are faced with a strongly acid soil, and know that rhododendrons will grow, then you can broaden your planting ideas to include other plants in the same family (*Ericaceae*), such as heather, mountain laurel, leatherleaf, pieris, blueberries, and others. If you are designing with plants, you may know that all plants in a particular family share certain features, which enables you to mix displays effectively and extend your range.

It is impossible to take in the entire plant kingdom in just one sitting. As a starting point you must approach one plant at a time. Fortunately, there is a kind of logic to the plant world that will save you from having to learn absolutely every plant. This is why the family groupings exist. With experience, it is possible to make sense of the enormous biological diversity of the plant kingdom, by piecing together family likenesses and genealogical connections.

Erica carnea,
winter heath

Vaccinium oxycoccos,
cranberry

Heathers and cranberries, together with rhododendrons, are in the *Ericaceae* family. Most members of this family prefer acidic soil and will not thrive on alkaline soils.

How to use this book

RECOGNIZE AND IDENTIFY

Using this book you can teach yourself to recognize similar characteristics among thousands of plants and identify which family they belong to. The families are listed in roughly evolutionary order, as they appear on the family tree (see pages 10–11). As you become confident with family characteristics, you will begin to recognize individual members in a family and see not only how they are related, but also how they differ.

MAJOR FAMILIES

Several of the major families, such as the grasses (*Poaceae*), the pines (*Pinaceae*), the buttercups (*Ranunculaceae*), and the beans (*Fabaceae*) are given a four-page treatment. It makes sense for beginners to familiarize themselves with these important groups first because they are relatively easy to recognize and encompass a good number of commonly cultivated and wild plants.

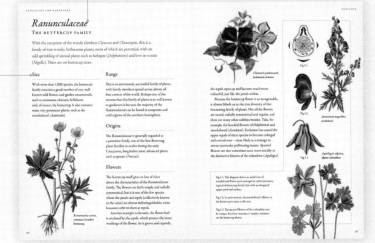

KEY FACTUAL DETAILS

Each chapter includes sections covering the size of the family, its range and origins, as well as key details on the commonalities and variations of leaf and flower formation within the family.

DIAGRAMS

Beautiful and informative illustrations and diagrams help to aid identification.

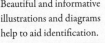

USES IN THE GARDEN

Text boxes in most chapters give helpful insights into how to use and get the most out of plants from each family in the garden.

CLASSICAL AND CONTEMPORARY RESEARCH

The genealogy of plants is a complicated subject. Experts disagree on many lineages and new evidence in the form of DNA analysis presents all sorts of new questions. This book combines classical and contemporary thinking that will bring plant genealogy alive and make it relevant to the modern gardener or naturalist.

Introducing plant family trees

For centuries botanists have worked to describe and explain the great diversity of plants. Similarities were noted while comparing plant leaves, stems, flowers, and other structures. These commonalities were used to group together plants, thereby forming a basic classification. Over time this was refined to include several hierarchical layers, including families. Traditional classifications were used primarily to identify plant species, but with the discovery of what is now known as genetics, it became clear that these similarities were inherited and their presence implied a relationship. Today's plant classifications aim to reflect these genetic relationships between plant species, often represented using branching diagrams that resemble the family trees of genealogists, such as on the following page.

Acer palmatum,
Japanese maple

Aesculus hippocastanum,
European horse chestnut

Building family trees

Plant genealogies are created by comparing data from numerous plant groups; pairs that share the most similarities are likely to be closely related. Several types of data can be used but most modern studies rely on DNA. As in all organisms, long strands of DNA (or deoxyribonucleic acid) form chromosomes, which are an instruction manual for the creation of the organism. The full complement of DNA in each cell, known as the genome, can be vast. The perennial herb *Paris japonica* (*Melanthiaceae*) has the largest known genome of any plant, at around 150 billion base pairs—compare that with the human genome, which has a measly 3.2 billion base pairs! Analyzing such large amounts of data is difficult, so botanists select a few variable regions and compare these across multiple species. A family tree is built by linking together those plants whose DNA regions are most similar. Subsequent branches are added for species with fewer genetic similarities, the whole process being operated by computer algorithms.

What do we learn from family trees?

Plant family trees, such as the one on the following page, are created using DNA and other data. They reveal much about plant relationships. To know how different species and families are related confirms how accurately they have been classified. For example, such family trees show that, according to their DNA, maples and horse chestnuts belong in the soapberry family (*Sapindaceae*), instead of residing in their former families (*Aceraceae*, *Hippocastanaceae* respectively).

By examining the positions of different families on the tree we can see how the physical characteristics of a plant have changed over time. The family tree on the following page illustrates the evolutionary development of the plant kingdom, from the earliest-known plant groups to the modern plant families we know today.

The genealogy of plants, illustrated by family trees, is primarily a tool for identification. Understanding both the physical characteristics that distinguish the major plants groups and how they developed allows a gardener to accurately identify any plant. Moreover, newer classifications, developed using modern DNA-based methods, are the most accurate ever developed, ensuring that the families described in this book are much less likely to change in the future.

Hydrangea macrophylla, bigleaf hydrangea

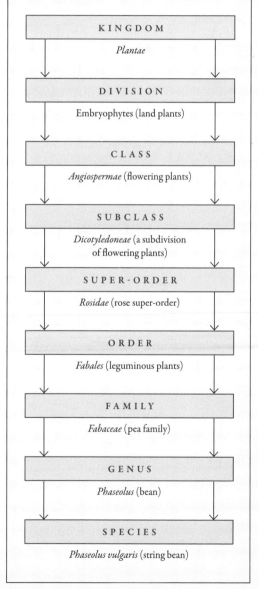

SCIENTIFIC NOMENCLATURE

The application of formal scientific names is governed by an international code of nomenclature. These are the main units used in plant nomenclature, using the string bean as an example:

KINGDOM
Plantae

DIVISION
Embryophytes (land plants)

CLASS
Angiospermae (flowering plants)

SUBCLASS
Dicotyledoneae (a subdivision of flowering plants)

SUPER-ORDER
Rosidae (rose super-order)

ORDER
Fabales (leguminous plants)

FAMILY
Fabaceae (pea family)

GENUS
Phaseolus (bean)

SPECIES
Phaseolus vulgaris (string bean)

The plant family tree

This tree was created by comparing DNA samples from a range of plants, looking for genetic similarities and differences. Progressing from left to right, with the earliest plant groups farthest left, it indicates the progress of evolution from now-extinct ancestors to the modern plant families we know today (farthest right). Like a human family tree, families that are closely related are close to one another, while those most distantly related appear farthest apart.

To reduce the complexity of this plant family tree, we've included only those families described in this book. Each branch on this tree represents the appearance of a new group: first liverworts and mosses (bryophytes), then ferns.

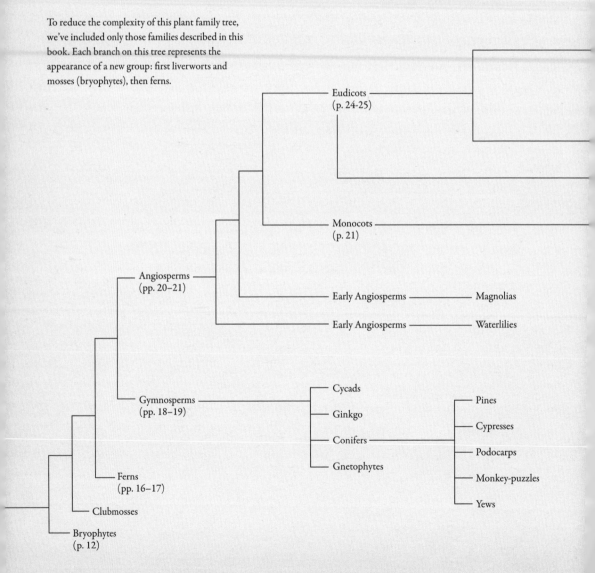

Eudicots (p. 24-25)

Monocots (p. 21)

Angiosperms (pp. 20–21)

Early Angiosperms — Magnolias

Early Angiosperms — Waterlilies

Gymnosperms (pp. 18–19)

Cycads

Ginkgo

Conifers

Gnetophytes

Pines

Cypresses

Podocarps

Monkey-puzzles

Yews

Ferns (pp. 16–17)

Clubmosses

Bryophytes (p. 12)

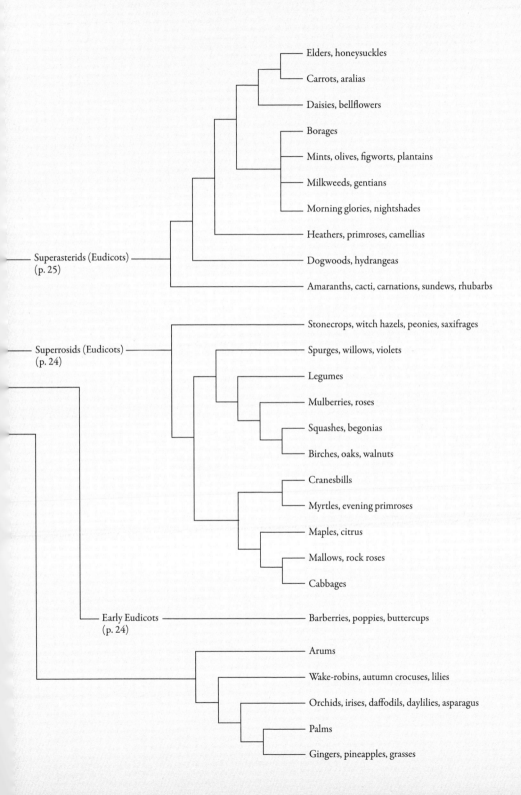

From the first plants to flowers

The plants we see around us today, like all life on Earth, have their origins in the ocean. The transition to life on land probably began around 500 million years ago, but from these early land plants, which likely resembled green algae, there were to evolve a group that now dominates the Earth. The flowering plants, or angiosperms, that today provide most of our garden plants, are the culmination of several evolutionary leaps. Understanding these major developments is crucial for identifying the major groups of garden plants.

Avoiding drought

The primary limiting factor for the first plants moving onto dry land was water. Early plants lacked roots and were unable to actively transport water, so were limited to permanently wet areas. Some of these early lineages gave rise to liverworts, hornworts, and mosses, which today thrive in damp habitats—most noticeable in the yard around pond edges, on shady lawns, and on the surface of potting compost. Collectively known as bryophytes, these plants lack real roots and an effective vascular system; instead, all parts of bryophytes can absorb moisture directly. Modern bryophytes can only grow when moisture is abundant, and survive drought by drying themselves out almost entirely.

Marchantia polymorpha,
umbrella liverwort

Equisetum sylvaticum,
wood horsetail

Selaginella martensii (little clubmoss) produces trailing stems and is a good ground cover plant for moist, shady places.

The first plants to be able to transport water from their roots through their stems appeared 430 million years ago. This vascular system was a great evolutionary leap that countered the problem of access to water. Not only could these plants colonize and grow in drier habitats, but they were also able to grow larger than the diminutive bryophytes. Plants could keep their tissues firm and upright because the vascular system replenished the cells when water was lost to drought; larger plants could withstand moderate desiccation without wilting.

The development of a vascular system led to the first woody plants, and around 380 million years ago the land was dominated by giant trees up to 165 feet tall. These trees would ultimately go extinct, leaving behind a few modern relations, including clubmosses (*Lycopodiaceae*) and horsetail (*Equisetaceae*), the scourge of perennial borders.

This reconstruction of the tree *Calamites* (pictured right) is based on fossil remains. It is an extinct relative of modern horsetails (*Equisetum*).

Modern leaves

Another hugely significant evolutionary milestone was the advent of the modern leaf. Bryophytes and clubmosses have either no leaves or small, simple ones with a single vein down the center. The first trees bore these basic leaves; known as microphylls, they could reach 3 feet in length.

During the period around 400 million years ago, plants with forked stems developed webbing between the stems, thereby forming the first modern leaf (known as a megaphyll). This more complex leaf was well supported and could therefore grow much larger than the basic microphylls. It was also better endowed with vascular tissue, enabling it to remain hydrated. It is thought that during this period, megaphylls arose independently a number of times, in several different plant groups. The ferns, which first appeared around 360 million years ago, are descendants of one of these groups.

Treelike plant growing from underground stem, known as a rhizome

Reproduction

While a vascular system combined with megaphylls allowed plants some freedom to disperse away from moist habitats, one feature continued to constrain their ability to spread: reproduction. Ferns perpetuate by releasing spores from fertile sites on the leaves (known as sori). These spores drift on the wind before settling to earth where they germinate, forming a green, fleshy structure known as a prothallus. Resembling some liverworts, the prothallus produces sperm that swim across the ground to fertilize the egg cells of another prothallus. Of course, sperm can only swim in moist conditions, but this problem was alleviated by the development of two kinds of spore of different sizes. Large spores were retained within structures on the leaves and germinated *in situ*, thus protecting the prothallus from desiccation. Small spores were windborne, and germinated only upon reaching the large spore,

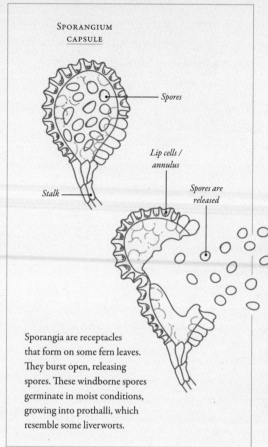

SPORANGIUM CAPSULE

Spores

Lip cells / annulus

Stalk

Spores are released

Sporangia are receptacles that form on some fern leaves. They burst open, releasing spores. These windborne spores germinate in moist conditions, growing into prothalli, which resemble some liverworts.

where the sperm could swim within the protected structure. Once fertilized, the large spore and its surrounding protective layers formed a seed. The first seed plants appeared around 350 million years ago. Seeds arose from the leaves, which over time were reduced to scales that formed cones. Gymnosperms, which include familiar garden plants such as conifers, produce seeds in cones.

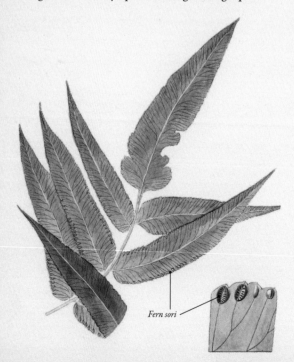

Fern sori

Marattia laxa,
potato fern
Ferns reproduce by releasing dustlike spores from structures called sporangia, borne on their foliage. Sporangia may be clustered together forming discrete sori (as seen here) or evenly spread across the leaf underside.

Cones, such as this pine cone from *Pinus gerardiana*, are produced by many gymnosperms to protect their seeds. The scales that bear the seeds are thought to derive from spore-bearing leaves.

Flowers

Ferns, gymnosperms, and many other now-extinct groups dominated the land for more than 200 million years, establishing an environment for the rise of the dinosaurs. However, one significant plant group had yet to emerge. The first flowering plants appeared abruptly, according to fossil records, around 130 million years ago. The evolution of the flower is the result of the large and small spore-producing tissue coming together into one unit, and represents the final milestone in the development of modern plants. Leaves that held large spores evolved to protect them from damage by wrapping around them. The resulting structure is now known as an ovary. Leaves that produced small spores gradually reduced in size to form stamens. Further protection was provided by surrounding the ovary and stamens with leaves, which over time became petals and sepals.

The sudden appearance of angiosperms (flowering plants, see pages 20–22) in the Cretaceous has long puzzled botanists. Evolution, as envisaged by Darwin, involves the slow change and development of physical structures over multiple generations; Darwin described the comparatively rapid appearance of flowering plants as an "abominable mystery." He theorized that angiosperms evolved much earlier on a now-vanished continent in the southern hemisphere, suggesting that the lack of early fossils was a result merely of an insufficiently known fossil record. A modern view is that evolution does not proceed at a steady pace; the appearance of the angiosperms coincided with a diversification in insects that pollinated the early flowers. Insects facilitated a rapid evolution of angiosperm families by actively spreading the genetic material in their pollen.

Whatever it was that triggered their appearance, the success of flowers is demonstrated by the fact that today around ninety-five percent of all vascular plants are angiosperms; they make up most of our garden flora.

***Illicium anisatum*,**
Japanese star anise
In most flowers, green sepals protect the bud while colorful petals attract pollinators. However, in star anise, as in many early angiosperms including waterlilies and magnolias, the sepals and petals appear similar.

Flower in cross-section　　*Petals removed to reveal stamens and carpels*　　*Fruit and seed*

Ferns—fronds to fiddleheads

Most gardeners will have their own mental image of what a fern looks like; triangular leaves, often deeply divided, probably resembling common bracken (*Pteridium aquilinum*). But what exactly is a fern? Plant taxonomists are currently grappling with that very question. Ferns are fundamentally vascular plants with leaves, known as fronds, that reproduce by releasing dustlike spores.

Not all ferns resemble bracken; the horsetails (*Equisetum*), whose DNA indicates placement within this ancient assemblage, are a prime example of fern diversity. They ally with several lineages, including the tropical giant ferns (*Marattiaceae*), leafless whisk ferns (*Psilotaceae*), and small adder's tongues (*Ophioglossaceae*). These early groups account for only 3 percent of all fern species; the remainder, the modern ferns, are the most useful in the garden.

Pteridium aquilinum, **common bracken**

Forest ferns

While ferns first appeared in the Late Devonian (360 million years ago), modern ferns are of a much more recent vintage, with recognizable groups appearing from around the end of the Jurassic (150 million years ago). As flowering plants spread around the world, ferns found a niche growing under the canopy of the angiosperm forests. They underwent a period of great diversification, and today they can be found in a variety of habitats, including under water, in exposed alpine areas, and on arid deserts.

There are around 9,100 species of modern ferns. One of their characteristic features is the "fiddlehead," which resembles both the head of a

Maidenhair spleenwort (*Asplenium trichomanes*, left) and black spleenwort (*A. nigrum*, right), like all members of the family *Aspleniaceae*, release spores from linear or rod-shaped sori on the frond undersides.

violin and a shepherd's crook or crozier, and contains the furled fronds of the young fern. Fern fronds are hugely variable, ranging from completely undivided to lacy and finely divided. Spores are released from structures called sporangia that can usually be found on the lower underside of the frond. Fern sporangia are clustered together to form a sorus (pl. sori), though a few ferns have sporangia spread in a uniform mass across the leaf's underside. Fronds arise from a rootlike stem known as a rhizome, which can be short and squat or elongated. In tree ferns, this stem forms an upright trunk, while in many other species it grows horizontally across (or under) the ground or attaches to a tree branch.

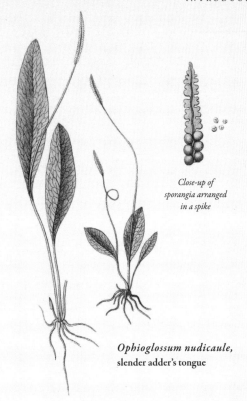

Close-up of sporangia arranged in a spike

Ophioglossum nudicaule,
slender adder's tongue

Dryopteris filix-mas,
male fern

The leaves of ferns, known as fronds, unravel from coiled fiddleheads. As in most members of the family *Dryopteridaceae*, the sori are circular and protected by an umbrella-shaped structure called the indusium.

Spore-bearing sori

Sorus in section with umbrella-like membranous covering (indusium)

USES FOR FERNS

Ferns do not have a chapter of their own in this book because their classification remains in a state of flux. Of course, many ferns are in their element in woodland gardens, especially those in the genera *Adiantum*, *Dryopteris*, and *Polystichum*, making them extremely useful plants to grow in those difficult, dry, shady spots under trees. Plant the vigorous sensitive and ostrich ferns (*Onoclea sensibilis* and *Matteuccia struthiopteris*) in damp soil. Many spleenworts (*Asplenium*) can be tucked into cracks in walls and paving. Stately tree ferns (*Dicksonia*) give any yard an exotic feel, though they're only suited to milder climates. Indoors, the dramatic antler-like staghorn (*Platycerium*) has great presence. Partial shade and regular watering are essential for most ferns to thrive in the yard, but their bold foliage is readily incorporated into numerous garden settings.

Gymnosperms

The earliest seed plants still alive today are known as gymnosperms, a name that translates as "naked seeds." Unlike in flowering plants, whose seeds are enclosed within a fruit, gymnosperm seeds develop unenclosed, though often attached to the scales of a cone. Gymnosperms were the dominant plant group on land for many millions of years and, though they have now been superseded by the flowering plants, they are still important components of forests around the world.

Take the conifers, the largest group of living gymnosperms; they hold the record for the world's tallest tree (*Sequoia sempervirens*, coast redwood), largest tree by volume (*Sequoiadendron giganteum*, giant redwood), stoutest tree (*Taxodium mucronatum*, Montezuma cypress), and oldest tree (*Pinus longaeva*, Great Basin bristlecone pine). The coniferous forests of the Taiga stretch around the world from Alaska to Siberia, interrupted only by the Atlantic Ocean; it is the world's largest single biological community after the oceans.

GYMNOSPERM
OVULE

Micropyle

Integuments

Megaspore (egg cell)

Nucellus

Unlike in flowering plants (see page 20), gymnosperm ovules are not enclosed within an ovary.

The cone bearers

Conifers comprise six plant families distributed almost worldwide. They are typically woody shrubs or trees with needlelike or scalelike leaves, though many tropical species have abandoned needles and scales for lush foliage. Most are evergreen with a few deciduous exceptions, such as larches and bald cypresses. Conifer cones are woody or leathery with seeds forming on the scales, later to be released when the cone opens. The families *Taxaceae* and *Podocarpaceae* have cones that are greatly reduced, with the scales becoming fleshy, berrylike structures. Of the six coniferous families, only *Sciadopityaceae* is not described in this book; it comprises a single species: the Japanese umbrella pine (*Sciadopitys verticillata*), also known as koyamaki.

Sequoiadendron giganteum, giant redwood
This huge tree from California can grow to almost 328 feet high and over 100 feet in girth at the base.

Ginkgo biloba,
maidenhair tree

Cycas rumphii,
queen sago
With its palmlike leaves removed,
the pollen-producing cone of a male
cycad (right) is easy to see.

Some studies suggest that the gnetophytes
are the closest living relations to
flowering plants, while others link
them more closely to conifers.

Though only a remnant remains
of a much larger and more diverse
group, the modern gymnosperms
are successful survivors. The conifers
in particular often make excellent
garden plants that are easy to
maintain, conspicuous in the
landscape, and long-living.

Other gymnosperms

The second largest of the cone-bearing clans is the
cycads. It is hard to associate these palmlike trees
with conifers, but they typically produce their
seeds in cones that form in the center of the crown
of leaves. The popular *Ginkgo biloba* (maidenhair
tree) is the only surviving member of the third
gymnosperm lineage. Widely distributed in the
past, the lonely ginkgo is reduced to a single
species that may be extinct in the wild. The fourth
and final gymnosperm group, the gnetophytes,
comprises three disparate families of little or no
use in the garden. Sometimes known as joint firs,
Ephedra (*Ephedraceae*) are almost leafless desert
shrubs and the source of the drug ephedrine (used
in asthma medication). Tropical trees, shrubs, and
vines make up *Gnetaceae* (with a single
genus *Gnetum*). *Welwitschiaceae* includes
only a single, bizarre species in the Namib
Desert: *Welwitschia mirabilis* has only
two leaves that grow continuously
throughout its life, forming long, leathery
straps attached to a squat, woody trunk.

Sciadopitys verticillata,
umbrella pine

Welwitschia mirabilis,
tree tumbo
This strange African desert plant has only
two leaves, which grow continuously.

Angiosperms

Today, flowering plants produce most of the world's food and are the mainstay of gardens. Their dominance is due to the flower, a very adaptable organ that can cause plants to evolve, creating new species. For instance, if a handful of plants within one population undergo a change in flower colour, it is likely they will attract a new pollinator. The new pollinator might not visit the plants with the original flowers, so pollen, which contains the plant's genes, will no longer be transferred between these different-coloured plants and, ultimately, they will become different species. This process has led to the huge diversity we see today.

Floral diversity

The adaptation of flowers to individual pollination strategies has resulted in the huge variation we see in flowering plants today. Plants utilize diverse pollinators including insects, reptiles, birds, bats, primates, the wind, and even flowing water. Some have intricate relationships with an individual species of bee or butterfly, while others throw open their petals to a range of visiting animals.

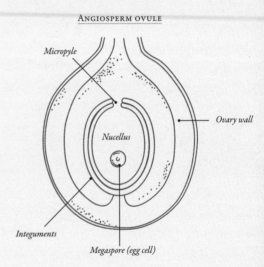

ANGIOSPERM OVULE

Micropyle

Ovary wall

Nucellus

Integuments

Megaspore (egg cell)

Flowering plant seeds develop within a protective ovary, which later becomes the fruit.

Petals and stamens
in multiples of three

Parallel
veins

Lilium maculatum,
Japanese orange lily

The world's largest flower, *Rafflesia arnoldii* of Indonesia, attracts carrion flies and beetles by releasing the scent of rotting meat from its 3-foot-wide blooms. At the other end of the scale, in the duckweed genus *Wolffia*, individual plants are less than ¹⁄₃₂ inch wide, and their correspondingly tiny flowers are the smallest in the world. These plants rarely bloom, and a pollinator has yet to be identified.

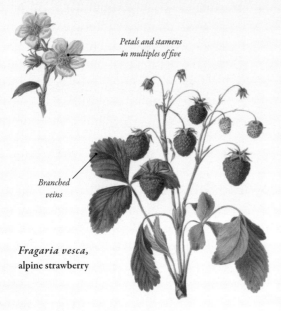

Petals and stamens in multiples of five

Branched veins

Fragaria vesca, alpine strawberry

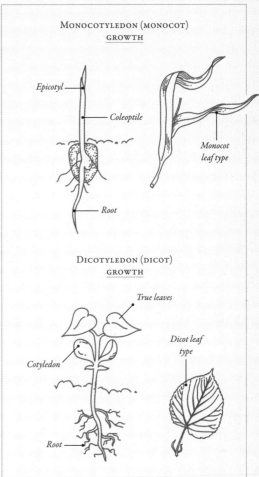

MONOCOTYLEDON (MONOCOT)
GROWTH

Epicotyl

Coleoptile

Monocot leaf type

Root

DICOTYLEDON (DICOT)
GROWTH

True leaves

Dicot leaf type

Cotyledon

Root

Monocots and dicots

Flowering plants were traditionally divided into two groups: monocots and dicots. Monocots, or monocotyledons to give them their full name, comprise about 15 percent of all angiosperms and include important families such as the grasses, palms, lilies, and orchids. The name is derived from the fact that their seeds each contain only one primary ("mono") embryonic leaf (cotyledon). Monocots typically have parallel veins in their leaves, and floral parts (petals, stamens, etc.) in multiples of three.

Most monocots are herbaceous, but there are a few treelike plants, such as palms, in which the vascular tissue is scattered throughout the stem. Dicots, unlike monocots, regularly produce trees, and their vascular tissue is arranged in rings, which allows tree age to be determined. Dicots, or dicotyledons, make up the remaining 85 percent of flowering plants. They have a pair ("di") of embryonic leaves in each seed; this trait is especially obvious at germination. Dicot leaves have branched veins and flowers that usually have floral parts in multiples of four or five.

Cotyledons are part of the developing embryo of the plant. In dicots, they are the first leaves to appear, but in monocots, they remain within the seed.

DNA-based studies have confirmed that monocots are a closely related group, derived from a single ancestor, but dicots are not. As a result, two groups (the ANA grade and Magnoliids, see page 22) have been removed from the dicots, leaving an abridged assemblage known as the eudicots (eudicotyledons). Even without the ANA grade and Magnoliids, which include garden plants such as waterlilies and magnolias, eudicots remain the largest group of plants on Earth and the foundation of our gardens.

Monocotyledons (monocots)

Farming began independently on several continents with the cultivation of cereal crops. The seeds of grasses such as wheat, rice, and maize were collected in the wild by humans, then gradually selected for size, quality, and ease of harvest. Today, the descendants of those wild grasses grow almost worldwide as cereal grasses (wheat, rice, maize, barley, and oats). Transported worldwide by humans for food, these monocotyledons (monocots) occupy billions of acres of prime habitat. Many other useful crops are harvested from monocots, including bananas, dates, coconuts, palm oil, sugarcane, ginger, and vanilla.

The ANA grade and Magnoliids

Putting monocots aside for a moment, it's worth mentioning a handful of precursor families that diverged before the monocots and eudicots. The ANA grade takes its name from three orders: *Amborellales*, *Nymphaeales*, and *Austrobaileyales*.

Nymphaea candida,
dwarf white waterlily

The first order includes only *Amborella trichopoda*, an obscure tree from New Caledonia, and likely the earliest branch on the angiosperm family tree. The second comprises waterlilies and their relatives, while the third is named for the tropical vine *Austrobaileya scandens* and includes star anise (*Illicium verum*) and magnolia vine (*Schisandra chinensis*).

Following these three orders are a diverse assemblage of 17 families called the Magnoliids. Well-known members include the magnolias (*Magnoliaceae*), laurels (*Lauraceae*),

Triticum aestivum,
bread wheat
First cultivated in the Middle East, bread wheat is now grown almost worldwide.

peppers (*Piperaceae*), and Dutchmen's pipes (*Aristolochiaceae*). While traditionally treated as dicots, many (the ANA grade and Magnoliids) also show characteristics of monocots. For example, waterlilies produce a single embryonic leaf from each seed, while custard apples (*Annonaceae*) and wild gingers (*Aristolochiaceae*) have flower parts in threes. Magnoliid flowers bear numerous tepals (petals and sepals that appear similar) and often have stamens with indistinct anthers and filaments.

Monocot evolution

Monocots have left behind a somewhat limited fossil record, largely because these primarily nonwoody plants are not readily preserved. Palms, which have tougher leaves and trunks than some other monocots, have been found as fossils, dating to around 90 million years ago, while fossilized

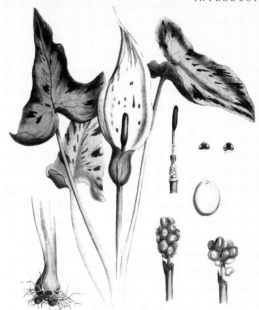

Arum maculatum,
lords-and-ladies

pollen from the arum family (*Araceae*) has been dated to between 120 and 110 million years ago. The very first branch on the monocot family tree is occupied by the small genus *Acorus* (*Acoraceae*), a semiaquatic rushlike herb also known as sweet flag. The next branch is similarly loaded with aquatics, such as flowering rush (*Butomaceae*), water-plantain (*Alismataceae*), eelgrass (*Zosteraceae*), and pondweed (*Aponogetonaceae*). This has led some botanists to theorize that monocots evolved from aquatic or semiaquatic ancestors, but fossil evidence is sparse.

Though accounting for less than a quarter of all flowering plants, monocots punch well above their weight in terms of economic importance. They also provide year-round color in the garden, from the blooms of daffodils, lilies, and orchids, to the textural foliage of grasses, hostas, and phormiums, and the bold structural presence of bananas, palms, and yuccas.

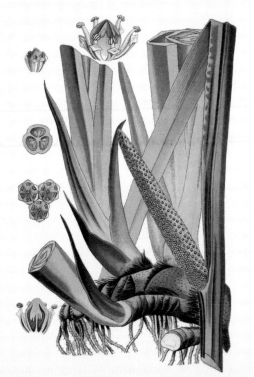

Acorus calamus,
sweet flag

Eudicotyledons (eudicots)

Scan any garden and most of the plants you will see are eudicots. They make up 75 percent of all angiosperms, dominate many terrestrial ecosystems, and provide important foods, fibers, and timbers. Of the five largest plant families on Earth, three are eudicots: *Asteraceae* (daisies, the largest), *Fabaceae* (legumes, third largest), and *Rubiaceae* (coffee, fourth largest). Given their great diversity, it can be difficult to characterize the group, though they usually have branched veins in their leaves and flower parts in fours or fives.
Great progress has been made in reconstructing the eudicot family tree and the classification of this important group has begun to stabilize.

As with monocots, the first branch on the tree is an aquatic plant. *Ceratophyllum demersum*, sometimes known as hornwort, is a waterweed familiar to some as an oxygenator for garden ponds and aquaria. It's the only genus in its family and has miniscule flowers, contrasting sharply with the roses, sunflowers, and other striking blooms that evolved from this common ancestor.

Leucanthemum vulgare, ox-eye daisy

Superrosids

Tracing the tree upward, the next few branches carry around 15 families known as basal eudicots, including the barberries (*Berberidaceae*), buttercups (*Ranunculaceae*), and poppies (*Papaveraceae*). These families often show characteristics common to earlier groups, such as multiple stamens and ovaries divided into separate carpels. From here the tree of life forks, with the remaining families (most of the eudicots) split between the two branches. One branch holds 119 families, a group known as the superrosids; the other carries 115 families, collectively known as superasterids. These two groups, named for a characteristic family (the rose and daisy

Beans, such as this *Phaseolus coccineus* (scarlet string bean), are members of the legume family *Fabaceae*, which includes peas, lupines, clover, and laburnum, and is the third largest plant family on Earth.

families, respectively) each contain around one-third of living angiosperms.

The superrosids contain many families of importance to gardeners, including roses (*Rosaceae*), legumes (*Fabaceae*), mallows (*Malvaceae*), birches (*Betulaceae*), begonias (*Begoniaceae*), and cranesbills (*Geraniaceae*). As with all large groups, it is difficult to find a definitive character for the superrosids because they are hugely variable. Perhaps the best is the presence of leaflike appendages known as stipules at the base of the petioles (the stalk that attaches the leaf blade to the stem), though some superrosids lack this feature, and it is found in some families outside this group (for example, *Rubiaceae*). Bear in mind that stipules can take many forms including glands, hairs, spines, or the more recognizable leafy types. It is not uncommon for stipules to be shed with age, so always inspect young growth.

Hibiscus mutabilis,
Confederate rose

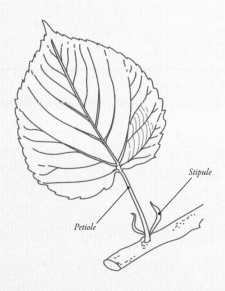

Stipules are leafy outgrowths at the base of the petiole. They often appear in pairs and can also take the form of thorns or glands.

Superasterids

The superasterids are even harder to define. They are split into two groups: one including families related to the carnations (*Caryophyllaceae*) and another containing the core asterids. *Caryophyllaceae* is allied to a diverse group of families that includes cacti (*Cactaceae*), amaranth (*Amaranthaceae*), rhubarb (*Polygonaceae*), and several carnivorous plants (*Droseraceae, Nepenthaceae*). They typically have leaves with entire (not lobed or toothed) margins (see page 33). The core asterids, on the other hand, mostly have tubular flowers with stamens reduced in number, often adhering to the inside of the petal tube. Horticulturally important families in this group include daisies (*Asteraceae*), borages (*Boraginaceae*), hydrangeas (*Hydrangeaceae*), mints (*Lamiaceae*), and honeysuckles (*Caprifoliaceae*).

What to look for when identifying plants

The intricacies and subtleties of plant identification are unfortunately beyond the reaches of a simple Internet search engine. The best we currently have to rely on are our own observational skills.

Gathering information

Gather as much evidence as you can, preferably by taking photographs or making detailed sketches. These days it is not necessary to dig up plants or break off pieces. If part of a plant must be taken away for identification you should first check that it is legal and that you have permission.

Survey the surrounding environment for clues. Is the plant in shade, by the sea, growing in water, or among taller or smaller plants? Is it in a woodland, by a road, in a disused quarry, or in someone's yard? In winter, fallen leaves and fruit are very helpful when identifying deciduous trees and shrubs.

Five-lobed leaves

Twining tendrils

Passiflora caerulea,
common passionflower

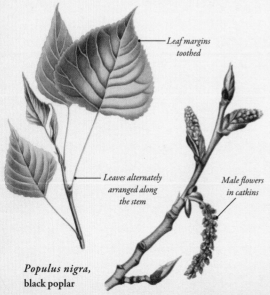

Leaf margins toothed

Leaves alternately arranged along the stem

Male flowers in catkins

Populus nigra,
black poplar

How to look at plants

At their most basic, plants are made up of roots, shoots, and leaves. You may also see flowers, catkins, cones, or fruit. Other details may include buds, hairs, tendrils, thorns, or peculiar roots. Remember also to use your other senses. Does the plant smell of anything? Is it rough or smooth?

You should never rely just on a single feature. This makes the job of identification much more challenging, and a solitary feature might not be representative of the species as a whole. For example, the leaf forms of some plants, such as ivies, can be highly variable.

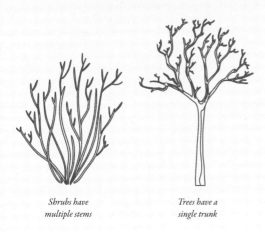

*Shrubs have
multiple stems*

*Trees have a
single trunk*

*Pinnate leaves with
many leaflets*

*Lemon-shaped
nuts*

The large pinnate leaves and walnut fruit on *Juglans cinerea* (white walnut) are key clues to this tree's identity, although fruits are only borne on female trees.

Allium karataviense (ornamental onion) is a short allium that is popular in gardens for its broad, grayish leaves and stout flower heads.

The key questions

Start with the obvious questions. Is the plant you are looking at woody, i.e. a tree or shrub? Is it aquatic, growing on or in water? Is it herbaceous (nonwoody), and if so, what size does it grow to? Is it one of the easily recognized groups, such as a fern, conifer, palm, orchid, or grass? Note that some plants are misleading; for example, not all "grasses" are grasses. Are you sure your specimen is even a plant? It could, for example, be a lichen or fungus.

Becoming familiar with plant anatomy

Leaves are not just leaves, and not all "flowers" are flowers. Compound leaves, for example, are made up of many leaflets, so what you are looking at might not be a single leaf, but one of several leaflets. What might appear at first to be a "flower" may actually be a flower head, made up of lots of smaller flowers. Flower heads might be a feathery grass plume, the spike of an arum flower, the flat-topped umbel of a carrot, or the composite head of a daisy flower. Take time to become familiar with the concepts listed over the next few pages.

*Rounded
flower head*

*Basal rosette
of strap-shaped,
arching leaves*

The different plant types

Plants can be characterized by family, by the length of their life cycle (annuals, biennials, and perennials), by their woodiness (woody or herbaceous), the type of habitat they occupy (water or desert), or the way they grow (bulbs or climbers). Many different plant "types" exist, and it is essential for gardeners to be familiar with these differences.

Annuals and perennials

Defined by the length of their life cycle, the main types of plant are ephemerals, annuals, biennials, and perennials. Ephemerals have extremely rapid life cycles and can germinate, grow, flower, set seed, and die back all within the space of a few weeks. They take advantage of short periods of favorable growth in otherwise hostile environments.

Lathyrus odoratus,
sweet pea
An annual climbing plant, grown anew each year from winter-sown seed and enjoyed for its pretty, scented flowers that are borne in summer.

Dianthus barbatus,
sweet william
With a limited life span of just two to three years, gardeners treat these plants as annuals or short-lived perennials.

Annuals and biennials have a slightly longer life cycle, completed inside one year for annuals, or two years for biennials. Sweet peas (*Lathyrus odoratus*) are popular annuals, and foxgloves (*Digitalis purpurea*) popular biennials. "Perennial" is a very broad term, covering any plant that lives for three years or more. It therefore covers woody and nonwoody plants and can be applied just as easily to a thousand-year-old tree as it can to a short-lived border plant.

Herbs, shrubs, and trees

To help refine the term "perennial," plants can be sorted to describe the way they grow. Thus, there are herbaceous (nonwoody) perennials, trees, shrubs, climbers and bulbs, corms, and tubers. Again, herbaceous perennial is another huge term, and can be further broken down into smaller arbitrary groupings, depending on requirements. These could be based on flowering time (late, mid, or early), preference for a certain soil (acidic or alkaline), light levels (sun or shade lovers), or family units (such as roses or palm trees)—the list could go on.

From seashore to mountaintop

Every habitat on Earth—almost—has its plant life. Thus, we have marsh plants, plants of coastal fringes, alpines, forest dwellers, aquatics, xerophytes (desert plants), and common or garden mesophytes (plants that grow in normal soil conditions with adequate moisture). There are many more plant types that could be described using this system. Understanding a plant's natural requirements is essential to successful planting in any garden.

Persicaria lapathifolia,
pale persicaria
This is an annual herbaceous plant that thrives on disturbed ground, such as cultivated soil, and can become weedy in some environments.

Tillandsia fasciculata,
cardinal airplant
Native to the tropical Americas, this is an epiphytic plant that grows without soil on the branches of trees in the rainforest. It collects all the water and nutrients it needs through its leaves from rain, dew, dust, and decaying plant and animal matter.

Geophytes, phanerophytes, epiphytes, and chamaephytes

Plants are grouped according to how they grow, using the Raunkiær system, devised by Christen C. Raunkiær and later extended by various authors. Geophytes are what gardeners call bulbs, corms, and tubers. Phanerophytes are plants that project their stems into the air: mostly trees, climbers, and shrubs. Epiphytes are those that grow on other plants, such as some orchids that grow on the branches of trees. Another important group is the chamaephytes; these are herbaceous perennials that, as part of their life cycle, periodically die back to resting buds at or near ground level.

Roots and stems

Plants use energy from the sun to create food in the form of carbohydrates, and draw water and minerals from the ground. To make this possible, their roots penetrate the soil, and stems and branches lift the leaves into the air where they are best placed to face the light. Roots also anchor the plant to the ground.

Roots

Trying to identify a live plant from its roots would be an impractical and environmentally insensitive thing to do. Nevertheless, it is sometimes possible to make certain assumptions without uprooting the plant and instead by taking a close look around the base of the plant. It helps to have a gardener's eye in order to do this.

Is the plant growing tidily in one place or spreading rapidly, like a weed? Don't be fooled by trees and shrubs, which are not always the single specimen they appear to be. For example, *Rhus typhina* (staghorn sumach) sends out a prolific number of suckering shoots from its roots. Herbaceous perennials often form clumps, and these can be tight (as seen in the genus *Helleborus*), or spread rapidly, forming new satellite clumps, as is the case with *Pycnanthemum muticum*. There is a gradient of behavior between these two.

Often the roots are visible above ground, as in the thickened, spreading rhizomes of *Geranium sanguineum*

Helleborus officinalis,
black hellebore

(bloody cranesbill)—although rhizomes are not roots but actually thickened, horizontal stems—or the ground-level stems of *Bergenia* (pigsqueak). Strawberry plants (*Fragaria*) have fibrous roots, but each summer after fruiting they send out long running shoots. The long underground runners of hedge woundwort (*Stachys sylvatica*) are found just under the soil surface, and they can run riot in garden borders. Plants growing from underground bulbs (sometimes corms or tubers) are quite easy to recognize. Their straplike leaves push up above the soil, usually in spring, then they flower and die back, all within a few months.

Mentha longifolia,
brook mint

*A creeping rhizome allows the
plant to spread rapidly*

Solanum tuberosum, **potato**

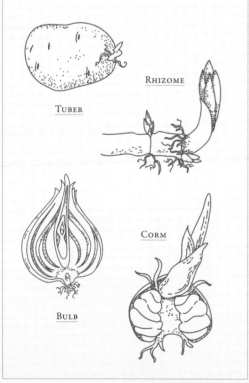

TUBER

RHIZOME

CORM

BULB

Stems

Plants do not always have stems, but at the very least there is often a structure supporting the flowers if not the leaves. The stems, trunk, or main branches form the principal structural element of a plant, and the branching structure can be extensive and intricate, or just made up of a number of upright or sprawling stems.

When seeking to identify a plant, consider whether the stem is woody or herbaceous. Stems, particularly if they are woody, will either be self-supporting, or twining or climbing. Look for modifications such as aerial roots, tendrils, spines, hooks, or thorns. If there is a clear trunk, make a note of any distinctive coloring, texture, or patterning. Bark can also come in a variety of colors and textures (peeling, flaking, splitting, ridged, etc.).

Herbaceous stems can be deciduous—dying back to a resting bud each year. Check the texture of the stem: is it hairy, rough, or smooth; round or angled? As a final measure, cut the stem to see if it is hollow or bleeds sap.

Rosa × centifolia, **Provence rose**

Leaves

The purpose of leaves is not only to harvest light energy from the sun, but also to allow the plant to "breathe." Carbon dioxide, oxygen, and water vapor all pass through tiny holes in the leaf surface in a process known as evapotranspiration. Leaves show many modifications to suit their environment.

As well as gas exchange and the capture of the sun's light, leaves also must contend with temperature extremes, humidity, and the attention of herbivores. The variety of leaf shapes and sizes, therefore, reflect these complex environmental interactions. There is such a huge variation between species that leaf form is a very useful tool in plant identification.

Pisum sativum,
garden pea
The leaves of peas are pinnate, with one or more pairs of leaflets and thin, twining tendrils that coil around any available support.

Simple or compound?

The first question to ask when looking at leaves is, are they simple or compound? This means, are the leaves just simple and undivided, or are they split into several or many smaller leaves, known as leaflets?

At their most basic, compound leaves can be bifoliate, with just two leaflets (though this

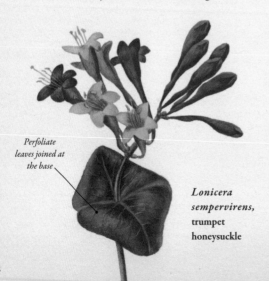

Perfoliate leaves joined at the base

Lonicera sempervirens, trumpet honeysuckle

is rare), or trifoliate, with three leaflets (which is more common, as seen in clover, *Trifolium*). Leaves can be palmately compound (like fingers on a hand) or pinnately arranged in feather-like pairs along an extension of the petiole (the stalk that connects the blade with the leaf base) known as a rachis. At the base of the petiole are stipules (see page 25). Sometimes the leaflets themselves are intricately divided into more leaflets.

Botanists use many specialist terms to describe leaf shapes. Simple leaves can be just as intricate, with deep divisions, serrations, or lobes, but they are never separated into individual leaflets. Examples include the straightforward elliptic leaf, such as a bay leaf (*Laurus*), the triangularly lobed leaves of ivy (*Hedera*), the perfoliate leaves of honeysuckle (*Lonicera*), and the needles of pines (*Pinus*).

The trifoliate leaves of *Trifolium repens* (white clover).

The palmately compound leaves of *Aesculus flava* (sweet buckeye).

The pinnately compound leaves of *Fraxinus americana* (white ash).

The ternately divided leaves of *Aquilegia vulgaris* (common columbine).

The rosetted rhombic leaves of *Micranthes hieracifolia* (hawk-leaved saxifrage).

Leaf margins

It is not just the leaf shape that is defined. The edge of the leaf (called the margin) is also extremely variable. Sometimes it is toothed or spiny, lobed or wavy, or it can be just smooth (described as an entire margin). Things can get quite complicated; toothed margins, for example, may be described as serrate (sawlike), serrulate (finely serrate), doubly serrate, dentate (toothed), or denticulate. The margins of leaflets on compound leaves must also be observed.

Alternate, opposite, or whorled?

Leaves are arranged on the plant to expose the surface of each leaf to light as efficiently as possible, without shading the plant's other leaves. It is one of the easiest features to look out for. The basic arrangement of leaves on a stem will usually fall into either opposite or alternate arrangement patterns. When three or more leaves are arranged in a circular pattern, they are known as being whorled. Sometimes leaves form a rosette at the base of the plant, as seen in foxglove (*Digitalis purpurea*), or at the end of a stem, as in houseleek tree (*Aeonium arboreum*).

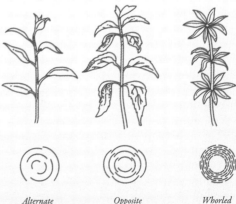

Alternate   *Opposite*   *Whorled*

Flowers

The plant kingdom can generally be split into flowering plants (angiosperms) and nonflowering plants (gymnosperms), although there are exceptions to these two groupings, such as ferns, for instance (see pages 16–17). Gymnosperms have cones for reproductive organs, instead of flowers. Only angiosperms have true flowers, and they are split into four parts: sepals, petals, stamens, and carpels.

The anatomy of a flower can be broken down into four constituent parts in concentric rings: outermost are the sepals and then the petals; innermost are the stamens and the carpel.

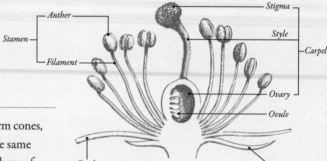

Flower structure

Flowers, which evolved from gymnosperm cones, share with the most primitive flowers the same whorled anatomical structure, such as those of *Magnolia* (pages 70–71). The four main whorls of a flower are referred to as the sepals (known collectively as a calyx), the petals (the corolla), the stamens (androecium), and the carpels (gynoecium), which are made up of the stigma, style, and ovary. Plants can have fewer, more specialized flower parts, as seen in the tiny flowers of grasses or flamboyant blooms of orchids.

Flower shapes

Close observation of flower anatomy is generally not necessary for recognizing the main family groups, although a little knowledge can certainly help. It is best to be able to recognise patterns in flower shape. For example, plants in the cabbage

Some plant families have very characteristic flower shapes. The daisy flowers of *Dahlia pinnata* (garden dahlia, far left) are unique to the *Asteraceae*, while the cruciform flowers of *Erysimum strictum* (wormseed wallflower, near left) typify the *Brassicaceae*, and the irregular and lipped flowers of *Ajuga reptans* (bugleweed, right) are typical of the mint family (*Lamiaceae*).

family (*Brassicaceae*) all have their four petals arranged in the shape of a cross, and the irregular (zygomorphic, or bilaterally symmetrical), lipped flowers of the mint family (*Lamiaceae*) are quite different from the radial symmetry of most regular (actinomorphic) flowers, such as a buttercup flower.

Inflorescences

Flowers are sometimes carried singly (on their own) but more often than not they are borne in flower heads, known as inflorescences. Like flower shapes, these too can be characteristic of certain families, such as the flat-topped umbels of the carrot family (*Apiaceae*) or the composite capitula of the daisy family (*Asteraceae*).

Inflorescences can be divided into cymes and racemes. Racemes tend to have a well-defined central stem, and there is usually no terminal flower at the end capable of further growth. The oldest flowers are near the bottom of the stalk. In a cyme, by contrast, every point of the flower head ends in a flower. Spikes, umbels, and daisy capitula are all technically classed as racemes, but have some differences. The major difference between

a spike and a raceme, for example, is that flowers on a spike lack pedicels (small stalks that bear individual flowers). Some flower heads are also branched. A panicle is an example of a raceme with many branches, each branch bearing a further raceme of flowers.

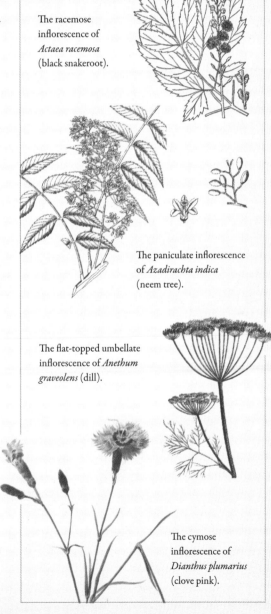

The racemose inflorescence of *Actaea racemosa* (black snakeroot).

The paniculate inflorescence of *Azadirachta indica* (neem tree).

The flat-topped umbellate inflorescence of *Anethum graveolens* (dill).

The cymose inflorescence of *Dianthus plumarius* (clove pink).

Fruit and seeds

The main purpose of a flower is to facilitate the transfer of pollen from anther to stigma, forming the fruit that carries the seeds of the next generation. In botanical parlance, a fruit is any seed-bearing structure, whether it is a pod, berry, nut, capsule, or acorn.

Inferior and superior ovaries

On many larger fruits, the remains of the flower can still be seen. On an apple, for example, these can be found at the end opposite the stalk. In tomatoes, the old calyx persists at the other, stalk end, and is often removed before the fruit is eaten. In each of these, the fruit, which has grown from the ovary, is either positioned below the calyx (where it is said to be inferior), or above it (superior). It is worth being aware of this, because plant families have flowers that are either inferior or superior. It can be, therefore, a defining feature. Hence flowers of apple trees (*Rosaceae*) have inferior ovaries, and those of tomatoes (*Solanaceae*) are superior.

Malus communis, apple

Seed

A swollen ovary forms a five-chambered pome fruit that contains seeds

Flower with petals (left) and without (right), showing the inferior position of the ovary

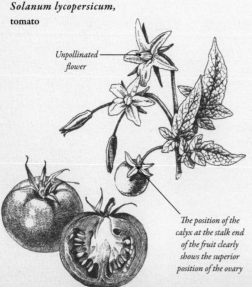

Solanum lycopersicum, tomato

Unpollinated flower

The position of the calyx at the stalk end of the fruit clearly shows the superior position of the ovary

Fruit

When they are ripe, fruits can be fleshy or dry. Dry fruits can be dehiscent, whereby they split open to release their seeds (such as a bean pod), or indehiscent, remaining closed (such as a hazelnut). Fleshy fruits can be simple, such as a tomato or rosehip, or multiple, resulting from the combined fruits of a single inflorescence. An example of a multiple fruit is mulberry (*Morus nigra*).

The strict definitions of fruit can seem a little bit arcane. A berry, for example, is a fleshy fruit that contains many seeds (such as a tomato, bell pepper, or gooseberry). However, an apple is not a berry but a pome, an orange is a hesperidium, and a peach is a drupe. Although it would be quite reasonable to assume that a drupe is a type of fleshy fruit, it is actually an indehiscent dry fruit surrounded by an outer fleshy layer. Examples of plants with drupes are plums, olives, and cherries.

The dried seed pods of *Albizia procera* split open to reveal the seeds within.

Seeds

Fruits are the means by which seeds are disseminated. Fleshy fruits tend to be eaten by animals, their seeds spread via droppings. Dry seeds are either eaten, carried by wind or water, or they attach themselves to passing animals.

There are a number of one-seeded dry fruits that are commonly referred to as seeds, such as the acorn (a nut), the cypsela of a sunflower (the kernel inside is actually the seed), and the seeds (achenes) of a strawberry.

Like leaves and flowers, many terms exist to describe seeds by their shape. Often they are self-explanatory, such as rounded, square, oblong, ovate, lenticular (lentil-shaped), and reniform (kidney-shaped). Seeds can also be distinctively colored, patterned, or textured.

Eudicotyledon seeds contain two seed leaves, which are usually round and fat because they retain the endosperm to feed the embryo plant. Monocotyledon seeds, in contrast, contain one seed leaf, which is thinner because the endosperm is kept separate.

Morus nigra,
black mulberry

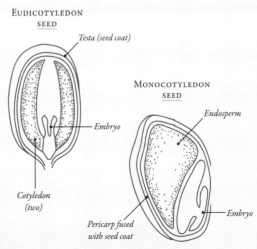

EUDICOTYLEDON
SEED

Testa (seed coat)

MONOCOTYLEDON
SEED

Endosperm

Embryo

Cotyledon
(two)

Pericarp fused
with seed coat

Embryo

Key to major groups

Identifying a mystery plant can seem difficult, but this series of keys should make the task a little easier. Starting from the top-left corner, answer each question, following the arrows until you reach the appropriate family. If you're uncertain how to answer, try both paths. Should you find that your plant matches more than one family, check each family account for further description. These keys are designed for use with common garden plants and only include families described in this book. If you can't find the right family, browse the individual accounts to look for similarities.

Key to major groups

This key will help you identify which major group your plant belongs to. Then, use the key for that group to identify your plant family. Some boxes can only be answered yes *or* no (not both); these have been included to provide additional identifying features.

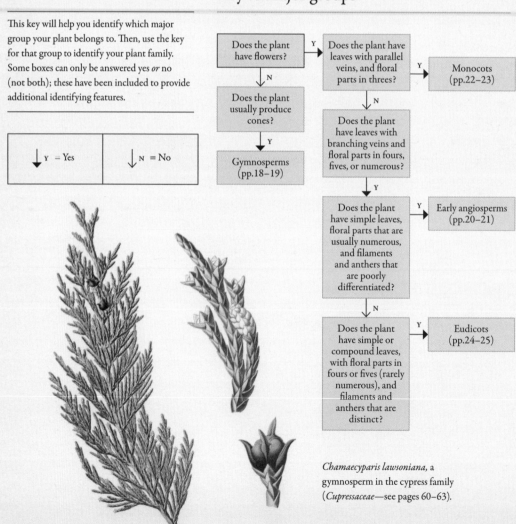

| ↓ Y – Yes | ↓ N = No |

Does the plant have flowers?

→ Y Does the plant have leaves with parallel veins, and floral parts in threes? → Y Monocots (pp.22–23)

↓ N

Does the plant usually produce cones?

↓ N Does the plant have leaves with branching veins and floral parts in fours, fives, or numerous?

↓ Y

Gymnosperms (pp.18–19)

Does the plant have simple leaves, floral parts that are usually numerous, and filaments and anthers that are poorly differentiated? → Y Early angiosperms (pp.20–21)

↓ N

Does the plant have simple or compound leaves, with floral parts in fours or fives (rarely numerous), and filaments and anthers that are distinct? → Y Eudicots (pp.24–25)

Chamaecyparis lawsoniana, a gymnosperm in the cypress family (*Cupressaceae*—see pages 60–63).

Key to gymnosperms

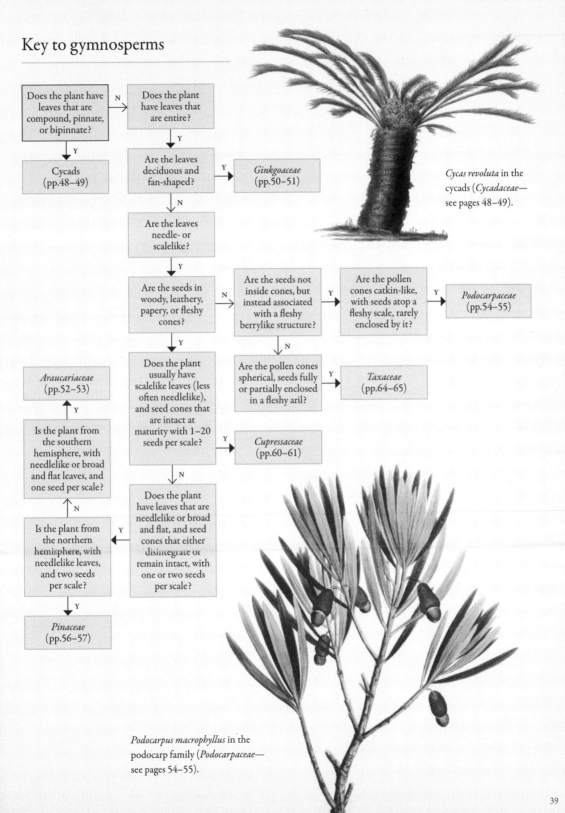

Does the plant have leaves that are compound, pinnate, or bipinnate? — N → Does the plant have leaves that are entire?

Does the plant have leaves that are compound, pinnate, or bipinnate? — Y → Cycads (pp.48–49)

Does the plant have leaves that are entire? — Y → Are the leaves deciduous and fan-shaped?

Are the leaves deciduous and fan-shaped? — Y → *Ginkgoaceae* (pp.50–51)

Are the leaves deciduous and fan-shaped? — N → Are the leaves needle- or scalelike?

Are the leaves needle- or scalelike? — Y → Are the seeds in woody, leathery, papery, or fleshy cones?

Are the seeds in woody, leathery, papery, or fleshy cones? — N → Are the seeds not inside cones, but instead associated with a fleshy berrylike structure?

Are the seeds not inside cones, but instead associated with a fleshy berrylike structure? — Y → Are the pollen cones catkin-like, with seeds atop a fleshy scale, rarely enclosed by it?

Are the pollen cones catkin-like, with seeds atop a fleshy scale, rarely enclosed by it? — Y → *Podocarpaceae* (pp.54–55)

Are the seeds not inside cones, but instead associated with a fleshy berrylike structure? — N → Are the pollen cones spherical, seeds fully or partially enclosed in a fleshy aril?

Are the pollen cones spherical, seeds fully or partially enclosed in a fleshy aril? — Y → *Taxaceae* (pp.64–65)

Are the seeds in woody, leathery, papery, or fleshy cones? — Y → Does the plant usually have scalelike leaves (less often needlelike), and seed cones that are intact at maturity with 1–20 seeds per scale?

Does the plant usually have scalelike leaves (less often needlelike), and seed cones that are intact at maturity with 1–20 seeds per scale? — Y → *Cupressaceae* (pp.60–61)

Does the plant usually have scalelike leaves (less often needlelike), and seed cones that are intact at maturity with 1–20 seeds per scale? — N → Does the plant have leaves that are needlelike or broad and flat, and seed cones that either disintegrate or remain intact, with one or two seeds per scale?

Does the plant have leaves that are needlelike or broad and flat, and seed cones that either disintegrate or remain intact, with one or two seeds per scale? → Is the plant from the northern hemisphere, with needlelike leaves, and two seeds per scale?

Is the plant from the northern hemisphere, with needlelike leaves, and two seeds per scale? — N → Is the plant from the southern hemisphere, with needlelike or broad and flat leaves, and one seed per scale?

Is the plant from the southern hemisphere, with needlelike or broad and flat leaves, and one seed per scale? — Y → *Araucariaceae* (pp.52–53)

Is the plant from the northern hemisphere, with needlelike leaves, and two seeds per scale? — Y → *Pinaceae* (pp.56–57)

Cycas revoluta in the cycads (*Cycadaceae*— see pages 48–49).

Podocarpus macrophyllus in the podocarp family (*Podocarpaceae*— see pages 54–55).

Key to early angiosperms and monocots

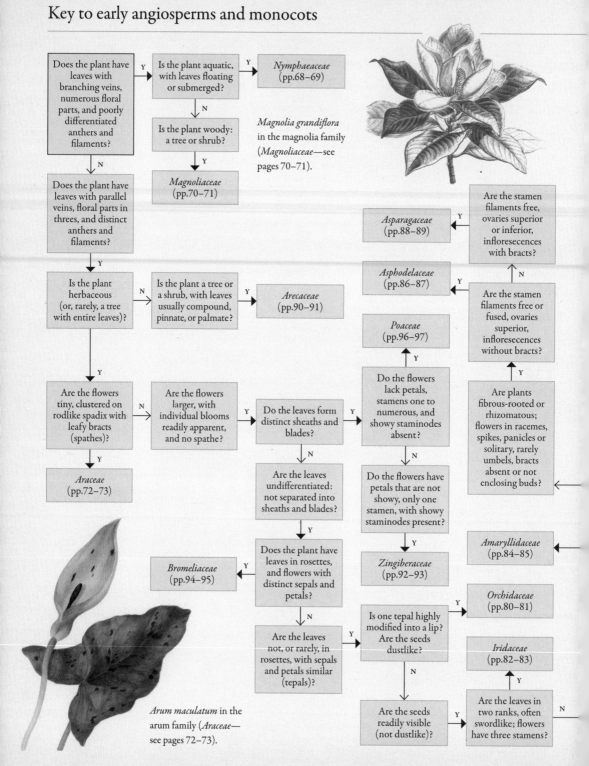

Does the plant have leaves with branching veins, numerous floral parts, and poorly differentiated anthers and filaments? **Y →** Is the plant aquatic, with leaves floating or submerged? **Y →** *Nymphaeaceae* (pp.68–69)

N ↓ Is the plant woody: a tree or shrub? **Y →** *Magnoliaceae* (pp.70–71)

Magnolia grandiflora in the magnolia family (*Magnoliaceae*—see pages 70–71).

N ↓ Does the plant have leaves with parallel veins, floral parts in threes, and distinct anthers and filaments?

Y ↓ Is the plant herbaceous (or, rarely, a tree with entire leaves)? **N →** Is the plant a tree or a shrub, with leaves usually compound, pinnate, or palmate? **Y →** *Arecaceae* (pp.90–91)

Y ↓ Are the flowers tiny, clustered on rodlike spadix with leafy bracts (spathes)? **N →** Are the flowers larger, with individual blooms readily apparent, and no spathe? **Y →** Do the leaves form distinct sheaths and blades? **Y →** Do the flowers lack petals, stamens one to numerous, and showy staminodes absent? **Y →** *Poaceae* (pp.96–97)

Y ↓ *Araceae* (pp.72–73)

N ↓ Are the leaves undifferentiated: not separated into sheaths and blades? **N ↓** Do the flowers have petals that are not showy, only one stamen, with showy staminodes present? **Y →** *Zingiberaceae* (pp.92–93)

Y ↓ Does the plant have leaves in rosettes, and flowers with distinct sepals and petals? **Y →** *Bromeliaceae* (pp.94–95)

N ↓ Are the leaves not, or rarely, in rosettes, with sepals and petals similar (tepals)? **Y →** Is one tepal highly modified into a lip? Are the seeds dustlike? **Y →** *Orchidaceae* (pp.80–81)

N ↓ Are the seeds readily visible (not dustlike)? **Y →** Are the leaves in two ranks, often swordlike; flowers have three stamens? **Y →** *Iridaceae* (pp.82–83)

N → Are plants fibrous-rooted or rhizomatous; flowers in racemes, spikes, panicles or solitary, rarely umbels, bracts absent or not enclosing buds? **←** (from Amaryllidaceae)

Amaryllidaceae (pp.84–85)

Are plants fibrous-rooted or rhizomatous; flowers in racemes, spikes, panicles or solitary, rarely umbels, bracts absent or not enclosing buds? **Y ↑** Are the stamen filaments free or fused, ovaries superior, inflorescences without bracts? **N →** Are the stamen filaments free, ovaries superior or inferior, inflorescences with bracts? **Y →** *Asparagaceae* (pp.88–89)

Y ↑ *Asphodelaceae* (pp.86–87)

Arum maculatum in the arum family (*Araceae*—see pages 72–73).

Crocus speciosus
in the iris family
(*Iridaceae*—see
pages 82–83).

Key to eudicot major groups

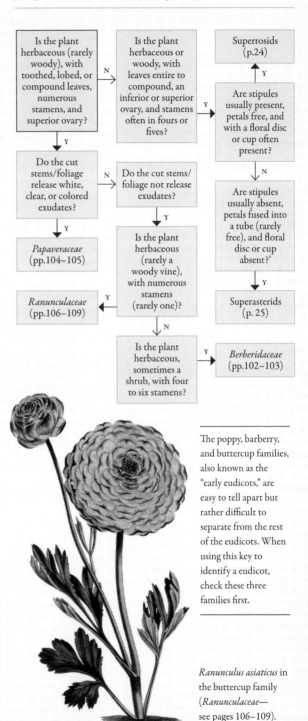

Is the plant herbaceous (rarely woody), with toothed, lobed, or compound leaves, numerous stamens, and superior ovary?

→ N →

Is the plant herbaceous or woody, with leaves entire to compound, an inferior or superior ovary, and stamens often in fours or fives?

→ Y →

Superrosids
(p.24)

↑ Y

Are stipules usually present, petals free, and with a floral disc or cup often present?

→ Y →

↓ N

Are stipules usually absent, petals fused into a tube (rarely free), and floral disc or cup absent?'

↓ Y

Superasterids
(p. 25)

↓ Y

Do the cut stems/foliage release white, clear, or colored exudates?

→ N →

Do the cut stems/foliage not release exudates?

↓ Y

Is the plant herbaceous (rarely a woody vine), with numerous stamens (rarely one)?

↓ Y

Papaveraceae
(pp.104–105)

Ranunculaceae
(pp.106–109)

← Y ←

↓ N

Is the plant herbaceous, sometimes a shrub, with four to six stamens?

→ Y →

Berberidaceae
(pp.102–103)

Colchicaceae
(pp.76–77)

↑ Y

Is the plant herbaceous or vining, with a corm, and leaves often sheathing the stems?

← N

Melanthiaceae
(pp.74–75)

↑ Y

Is the plant herbaceous, with rhizomes and leaves rarely sheathing the stems?

→ N →

← Y

Is the plant cormous or rhizomatous, with tepals unmarked, and anthers facing outward?

N ←

Are plants bulbous (rarely rhizomatous); flowers solitary or in umbels, with bracts enclosing flower buds?

Y ←

Liliaceae
(pp.78–79)

↑ Y

Is the plant bulbous or rhizomatous, with tepals often marked, and anthers facing inward?

← N

↑ Y

Are tepals often unmarked, with nectaries on ovary walls? Are the seeds shiny and black?

← N

Are the tepals often marked or patterned, with nectaries at the base of tepals or stamens? Are the seeds brown?

↑ Y

Are the leaves spirally arranged; flowers with six stamens?

→ Y →

The poppy, barberry, and buttercup families, also known as the "early eudicots," are easy to tell apart but rather difficult to separate from the rest of the eudicots. When using this key to identify a eudicot, check these three families first.

Ranunculus asiaticus in the buttercup family (*Ranunculaceae*— see pages 106–109).

Key to superrosids (woody plants)

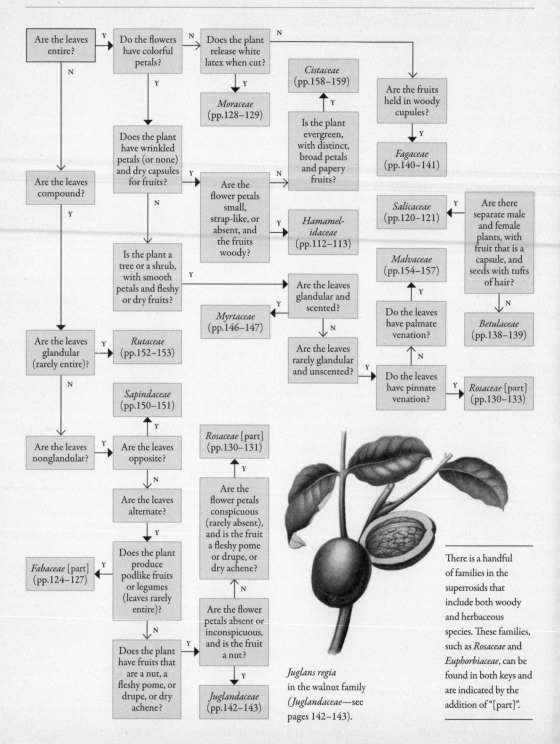

Are the leaves entire? — Y → Do the flowers have colorful petals? — N → Does the plant release white latex when cut? — N → Are the fruits held in woody cupules? — Y → *Fagaceae* (pp.140–141)

Does the plant release white latex when cut? — Y → *Moraceae* (pp.128–129)

Do the flowers have colorful petals? — Y → Does the plant have wrinkled petals (or none) and dry capsules for fruits? — Y → Are the flower petals small, strap-like, or absent, and the fruits woody? — N → Is the plant evergreen, with distinct, broad petals and papery fruits? — Y → *Cistaceae* (pp.158–159)

Are the flower petals small, strap-like, or absent, and the fruits woody? — Y → *Hamamel-idaceae* (pp.112–113)

Does the plant have wrinkled petals (or none) and dry capsules for fruits? — N → Is the plant a tree or a shrub, with smooth petals and fleshy or dry fruits? — Y → Are the leaves glandular and scented? — Y → *Myrtaceae* (pp.146–147)

Are the leaves glandular and scented? — N → Are the leaves rarely glandular and unscented? — Y → Do the leaves have pinnate venation? — Y → *Rosaceae* [part] (pp.130–133)

Do the leaves have pinnate venation? — N → Do the leaves have palmate venation? — Y → *Malvaceae* (pp.154–157)

Are there separate male and female plants, with fruit that is a capsule, and seeds with tufts of hair? — Y → *Salicaceae* (pp.120–121)

Are there separate male and female plants, with fruit that is a capsule, and seeds with tufts of hair? — N → *Betulaceae* (pp.138–139)

Are the leaves entire? — N → Are the leaves compound? — Y → Are the leaves glandular (rarely entire)? — Y → *Rutaceae* (pp.152–153)

Are the leaves glandular (rarely entire)? — N → Are the leaves nonglandular? — Y → Are the leaves opposite? — Y → *Sapindaceae* (pp.150–151)

Are the leaves opposite? — N → Are the leaves alternate? — Y → Does the plant produce podlike fruits or legumes (leaves rarely entire)? — Y → *Fabaceae* [part] (pp.124–127)

Does the plant produce podlike fruits or legumes (leaves rarely entire)? — N → Does the plant have fruits that are a nut, a fleshy pome, or drupe, or dry achene? — Y → Are the flower petals absent or inconspicuous, and is the fruit a nut? — Y → *Juglandaceae* (pp.142–143)

Are the flower petals absent or inconspicuous, and is the fruit a nut? — N → Are the flower petals conspicuous (rarely absent), and is the fruit a fleshy pome or drupe, or dry achene? — Y → *Rosaceae* [part] (pp.130–131)

Juglans regia in the walnut family (*Juglandaceae*—see pages 142–143).

There is a handful of families in the superrosids that include both woody and herbaceous species. These families, such as *Rosaceae* and *Euphorbiaceae*, can be found in both keys and are indicated by the addition of "[part]".

Key to superrosids (herbaceous plants)

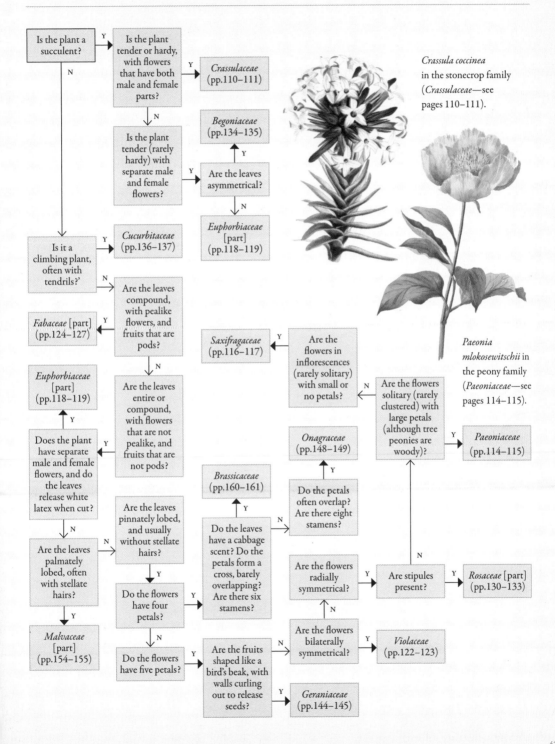

Is the plant a succulent?
- Y → Is the plant tender or hardy, with flowers that have both male and female parts?
  - Y → **Crassulaceae** (pp.110–111)
  - N → Is the plant tender (rarely hardy) with separate male and female flowers?
    - Y → Are the leaves asymmetrical?
      - Y → **Begoniaceae** (pp.134–135)
      - N → **Euphorbiaceae** [part] (pp.118–119)
- N → Is it a climbing plant, often with tendrils?'
  - Y → **Cucurbitaceae** (pp.136–137)
  - N → Are the leaves compound, with pealike flowers, and fruits that are pods?
    - Y → **Fabaceae** [part] (pp.124–127)
    - N → Are the leaves entire or compound, with flowers that are not pealike, and fruits that are not pods?
      - Y → Does the plant have separate male and female flowers, and do the leaves release white latex when cut?
        - Y → **Euphorbiaceae** [part] (pp.118–119)
        - N → Are the leaves palmately lobed, often with stellate hairs?
          - Y → **Malvaceae** [part] (pp.154–155)
          - N → Are the leaves pinnately lobed, and usually without stellate hairs?
            - Y → Do the flowers have four petals?
              - Y → Do the leaves have a cabbage scent? Do the petals form a cross, barely overlapping? Are there six stamens?
                - Y → **Brassicaceae** (pp.160–161)
                - N → Do the petals often overlap? Are there eight stamens?
                  - Y → **Onagraceae** (pp.148–149)
                  - N → Are the flowers in inflorescences (rarely solitary) with small or no petals?
                    - Y → **Saxifragaceae** (pp.116–117)
                    - N → Are the flowers solitary (rarely clustered) with large petals (although tree peonies are woody)?
                      - Y → **Paeoniaceae** (pp.114–115)
                      - N → Are the flowers radially symmetrical?
                        - Y → Are stipules present?
                          - Y → **Rosaceae** [part] (pp.130–133)
              - N → Do the flowers have five petals?
                - Y → Are the fruits shaped like a bird's beak, with walls curling out to release seeds?
                  - Y → **Geraniaceae** (pp.144–145)
                  - N → Are the flowers bilaterally symmetrical?
                    - Y → **Violaceae** (pp.122–123)

Crassula coccinea in the stonecrop family (*Crassulaceae*—see pages 110–111).

Paeonia mlokosewitschii in the peony family (*Paeoniaceae*—see pages 114–115).

Key to superasterids

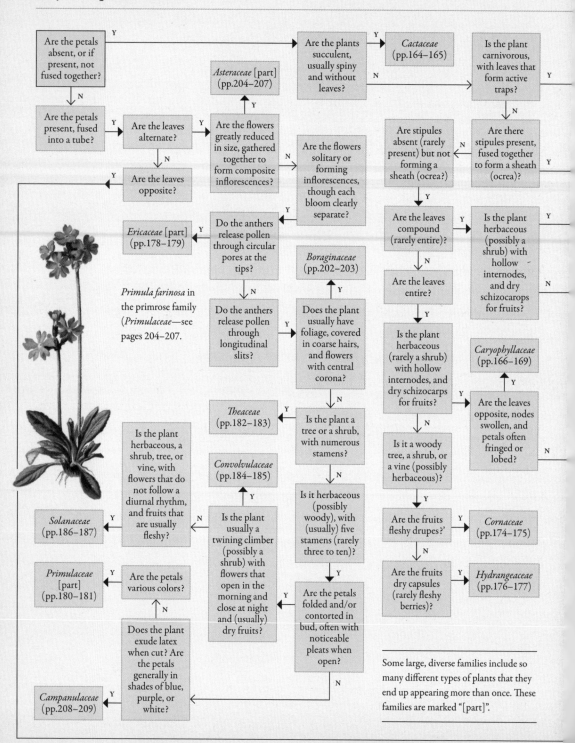

Are the petals absent, or if present, not fused together? — Y →

Are the petals present, fused into a tube? — Y →

Are the leaves alternate? — Y →

Are the leaves opposite? — Y ←

Are the flowers greatly reduced in size, gathered together to form composite inflorescences? — Y →

Asteraceae [part] (pp.204–207)

Are the flowers solitary or forming inflorescences, though each bloom clearly separate? — Y →

Do the anthers release pollen through circular pores at the tips? — Y →

Ericaceae [part] (pp.178–179)

Do the anthers release pollen through longitudinal slits? — Y →

Does the plant usually have foliage, covered in coarse hairs, and flowers with central corona? — Y →

Boraginaceae (pp.202–203)

Is the plant a tree or a shrub, with numerous stamens? — Y →

Theaceae (pp.182–183)

Is it herbaceous (possibly woody), with (usually) five stamens (rarely three to ten)? — Y →

Is the plant usually a twining climber (possibly a shrub) with flowers that open in the morning and close at night and (usually) dry fruits? — Y →

Convolvulaceae (pp.184–185)

Are the petals folded and/or contorted in bud, often with noticeable pleats when open?

Is the plant herbaceous, a shrub, tree, or vine, with flowers that do not follow a diurnal rhythm, and fruits that are usually fleshy? — Y →

Solanaceae (pp.186–187)

Are the petals various colors? — Y →

Primulaceae [part] (pp.180–181)

Does the plant exude latex when cut? Are the petals generally in shades of blue, purple, or white? — Y →

Campanulaceae (pp.208–209)

Are the plants succulent, usually spiny and without leaves? — Y →

Cactaceae (pp.164–165)

Is the plant carnivorous, with leaves that form active traps? — Y →

Are there stipules present, fused together to form a sheath (ocrea)? — Y →

Are stipules absent (rarely present) but not forming a sheath (ocrea)? — Y →

Are the leaves compound (rarely entire)? — Y →

Is the plant herbaceous (possibly a shrub) with hollow internodes, and dry schizocarps for fruits? — Y →

Are the leaves entire? — Y →

Is the plant herbaceous (rarely a shrub) with hollow internodes, and dry schizocarps for fruits? — Y →

Are the leaves opposite, nodes swollen, and petals often fringed or lobed? — Y →

Caryophyllaceae (pp.166–169)

Is it a woody tree, a shrub, or a vine (possibly herbaceous)? — Y →

Are the fruits fleshy drupes? — Y →

Cornaceae (pp.174–175)

Are the fruits dry capsules (rarely fleshy berries)? — Y →

Hydrangeaceae (pp.176–177)

Primula farinosa in the primrose family (*Primulaceae*—see pages 204–207.

Some large, diverse families include so many different types of plants that they end up appearing more than once. These families are marked "[part]".

44

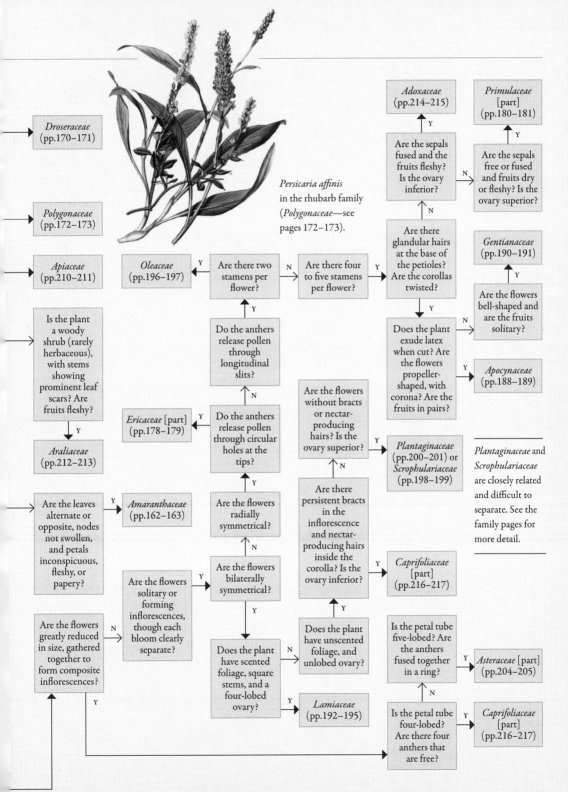

Persicaria affinis in the rhubarb family (*Polygonaceae*—see pages 172–173).

Droseraceae (pp.170–171)

Polygonaceae (pp.172–173)

Apiaceae (pp.210–211)

Is the plant a woody shrub (rarely herbaceous), with stems showing prominent leaf scars? Are fruits fleshy?

Araliaceae (pp.212–213)

Are the leaves alternate or opposite, nodes not swollen, and petals inconspicuous, fleshy, or papery?

Amaranthaceae (pp.162–163)

Are the flowers greatly reduced in size, gathered together to form composite inflorescences?

Are the flowers solitary or forming inflorescences, though each bloom clearly separate?

Oleaceae (pp.196–197)

Are there two stamens per flower?

Are there four to five stamens per flower?

Do the anthers release pollen through longitudinal slits?

Ericaceae [part] (pp.178–179)

Do the anthers release pollen through circular holes at the tips?

Are the flowers radially symmetrical?

Are the flowers bilaterally symmetrical?

Does the plant have scented foliage, square stems, and a four-lobed ovary?

Lamiaceae (pp.192–195)

Are the flowers without bracts or nectar-producing hairs? Is the ovary superior?

Are there persistent bracts in the inflorescence and nectar-producing hairs inside the corolla? Is the ovary inferior?

Does the plant have unscented foliage, and unlobed ovary?

Adoxaceae (pp.214–215)

Are the sepals fused and the fruits fleshy? Is the ovary inferior?

Are there glandular hairs at the base of the petioles? Are the corollas twisted?

Does the plant exude latex when cut? Are the flowers propeller-shaped, with corona? Are the fruits in pairs?

Primulaceae [part] (pp.180–181)

Are the sepals free or fused and fruits dry or fleshy? Is the ovary superior?

Gentianaceae (pp.190–191)

Are the flowers bell-shaped and are the fruits solitary?

Apocynaceae (pp.188–189)

Plantaginaceae (pp.200–201) or Scrophulariaceae (pp.198–199)

Plantaginaceae and *Scrophulariaceae* are closely related and difficult to separate. See the family pages for more detail.

Caprifoliaceae [part] (pp.216–217)

Is the petal tube five-lobed? Are the anthers fused together in a ring?

Asteraceae [part] (pp.204–205)

Is the petal tube four-lobed? Are there four anthers that are free?

Caprifoliaceae [part] (pp.216–217)

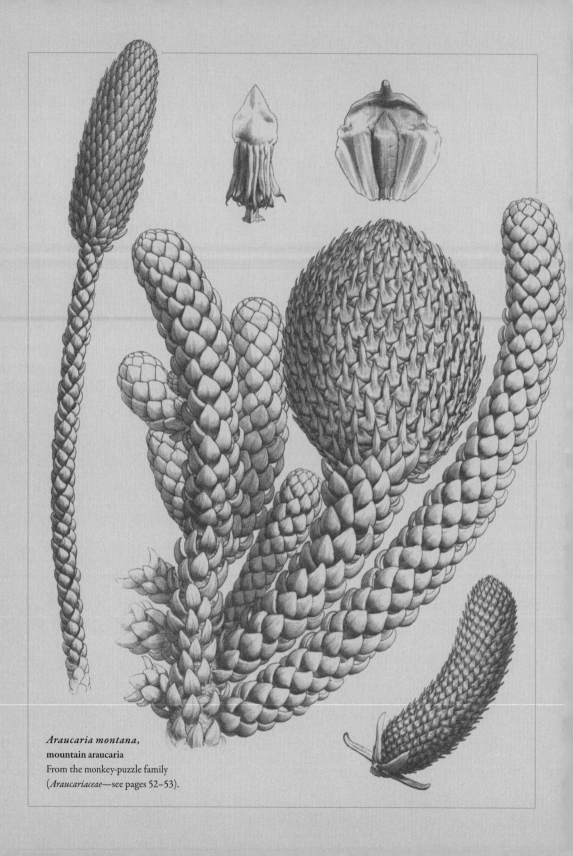

Araucaria montana,
mountain araucaria
From the monkey-puzzle family
(*Araucariaceae*—see pages 52–53).

CHAPTER 1

Gymnosperms

Gymnosperms, the earliest seed plants still alive today, are part of a larger group of plants called spermatophytes, which essentially includes all plants that produce seeds, such as the flowering plants (angiosperms). What distinguishes angiosperms from gymnosperms is the fact that gymnosperms do not produce flowers.

The term "gymnosperm" translates as "naked seed." This is another distinction between the two groups: in angiosperms the ovules are within an ovary; in gymnosperms they are not.

Gymnosperms are highly evolved plants, with complex vascular systems and specialized anatomical structures, such as woody tissue for support and cones for reproduction. They dominated the land for millions of years, and even today, though they have been superseded by flowering plants, gymnosperms still manage to occupy swathes of the Earth's surface.

Gardeners will find it easy to differentiate gymnosperms from all other plants, simply by their physical appearance; they include the familiar conifers and cycads.

Cycadaceae, Stangeriaceae, and Zamiaceae

THE CYCAD FAMILIES

This small group does not resemble the more familiar gymnosperms such as conifers, but rather could be mistaken for palms or ferns. They are typically single-stemmed trees that rarely form branches, or shrubs with almost subterranean trunks.

Size

There are 300 or so cycads, which are split between three families: *Cycadaceae* (*Cycas* only), *Stangeriaceae* (*Stangeria*, *Bowenia*), and *Zamiaceae* (eight genera).

Zamia poeppigiana,
Poeppig's zamia

Range

Cycads are generally restricted to the tropical regions of Africa, southeastern Asia, Australasia, and Latin America. Few species thrive in cooler climates; the widely cultivated Japanese sago (*Cycas revoluta*) is hardy in zones 8–11 but is often grown as a houseplant in cooler zones.

Origins

With fossils dating back to around 280 million years ago in the Permian and perhaps even earlier in the Carboniferous, cycads were around long before the dinosaurs walked the Earth. However, those that exist today are likely to have evolved in the last 12 million years.

Cones

Cycad cones appear in the center of the crown of leaves, with the male and female cones on separate plants. Male cones are composed of numerous scales, each with clusters of tissue that release large amounts of pollen. Female cones are similar, but often larger, more colorful, and with two ovules per scale. In the genus *Cycas*, ovules are produced on leaflike scales, loosely arranged at the stem tip, rather than in a true cone. They have two

*Cycas seeds form
leaflike scales*

Cycas circinalis,
East Indian sago
Cycad plants are either male or female. Males produce
pollen, while females, like this one, form seeds.

Endangered species

Given that they've survived for at least 300 million
years, you might think that nothing could threaten
the cycads, but at least a quarter of species are
endangered. The South African Wood's cycad
(*Encephalartos woodii*) is now extinct in the wild,
and known only from cuttings taken from the last
wild plant. Alas, the plant is male and no female
has ever been found, so Wood's cycad survives only
as these genetically identical cuttings in botanic
gardens. The most common threat to cycads is
habitat destruction, though illegal collection of
plants for horticulture is also problematic. Sago,
a starch produced from the trunk pith, which is
extracted for culinary purposes, can also threaten
cycad populations. Cycad sago is rich in
neurotoxins and must be carefully prepared to
prevent poisoning.

to eight ovules per scale. Cycads are usually
pollinated by beetles; the fertilized seeds often
develop colorful, fleshy coats, attracting the
animals that distribute them.

Leaves

The leaves are generally palmlike, with numerous
leaflets arranged along a central stalk. Uniquely,
Bowenia has twice-divided leaves. The foliage is
often tough and leathery, occasionally armed with
spines, and forms a crown at the stem tip. Old leaf
stalks line the trunk and scalelike leaves known as
cataphylls can be found among the regular foliage.
The veins on the leaves can be used to identify the
family: *Cycadaceae* leaflets have one central vein,
Stangeriaceae have one central vein with side
branches, and *Zamiaceae* have multiple veins
that run in parallel.

Bowenia spectabilis,
zamia fern

Ginkgoaceae

THE MAIDENHAIR TREE FAMILY

"Unique" is the best word to describe this deciduous tree. The wedge-shaped leaves distinguish it from any other known tree, and while the naked seeds place it in the gymnosperms, its exact relationship with plants in this group is still uncertain.

Size

There is only one living member of this family: *Ginkgo biloba*. The family is the only member of the order *Ginkgoales* and the class *Ginkgoopsida* and the division *Ginkgophyta*.

Range

Though widely cultivated around the world, the exact geographical origins of *Ginkgo biloba* are uncertain. Populations of this plant located in the Chinese province of Zhejiang were previously considered to be wild, but have since been found to harbor little genetic variation, suggesting they could have been planted. Ancient ginkgo trees at the edge of the Tibetan Plateau are more diverse, but given the popularity of their cultivation, it is difficult to know for sure whether or not ginkgo might in fact be extinct in the wild.

USES FOR THIS FAMILY

These ancient trees are surprisingly tolerant of pollution, so make a perfect street tree for urban areas. Planting females should be avoided, however, as the pungent smell of the seeds can be off-putting. In small yards, choose a dwarf cultivar such as "Gnome," "Troll," or "Mariken."

Origins

Ginkgo is famed for being a "living fossil." Ginkgo fossils are easily recognizable and date back to the Early Jurassic, 190 million years ago, with possible ancestral forms traced back 270 million years to the Permian. Such evidence can be found in fossil beds in many parts of the world, indicating that today's trees are relicts of a once more widespread group.

Ginkgo biloba,
maidenhair tree

Cones

As with cycads, ginkgos have separate male and female trees. The males produce catkin-like pollen cones as the new leaves emerge in spring, which hang down and allow the wind to disperse the pollen. Female trees do not produce cones, but rather pairs of ovules on long stalks, and each emits a sticky drop to capture windborne pollen. The ovules remain on the tree until the fall, when they develop a soft, fleshy outer coating. Fertilization occurs at this time and the fleshy coat begins to release a rather unpleasant smell. It is not clear which hungry animal is the target of this scent; it may by now be extinct.

Leaves

Ginkgo leaves resemble the leaflets of the fern genus *Adiantum*, also known as maidenhair ferns. They are fan-shaped and have a central cleft and sometimes additional splits or lobes. Plant breeders have developed great variety in ginkgo leaf forms, including those with variegated, tubular, or twisted leaves, those with extra clefts, and those with none. Before dropping in the fall, the leaves turn a vibrant butter-yellow color, making ginkgo a great choice to plant against dark evergreens. The tree itself has a rather sparse branching pattern, and other than on new growth, the leaves are clustered on top of short woody side shoots, rather like larches (*Larix*).

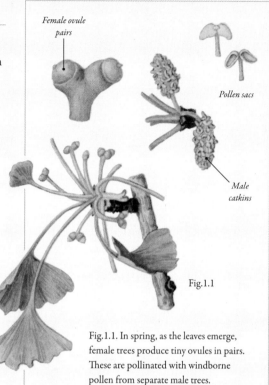

Female ovule pairs

Pollen sacs

Male catkins

Fig.1.1

Fig.1.1. In spring, as the leaves emerge, female trees produce tiny ovules in pairs. These are pollinated with windborne pollen from separate male trees.

Fig.1.2. In the fall, ginkgo leaves turn golden, while the ovules are fertilized using pollen stored from spring. The resulting seeds develop a fleshy outer coating.

Fig.1.2

Naked seed

Seed with fleshy coat

Araucariaceae

THE MONKEY-PUZZLE FAMILY

This very ancient coniferous group is made up of tall evergreen trees, and there is something distinctly primitive about all of them, with their thick columnar trunks and whorls of tiered branches.

Agathis dammara,
amboyna pine

In 1994, it was announced that there was a new addition to the family, Wollemi pine (*Wollemia nobilis*). Thought to have been extinct for more than 150 million years, a stand of these trees was found growing in a remote location in eastern Australia, and it is now being introduced into cultivation. Across the Tasman Sea in New Zealand exists the kauri (*Agathis australis*)—kauri trees are among the largest on the planet.

Range

What remains of this family is centered around the South Pacific, South America, and Australasia. The family name is taken from Arauco, a province of southern Chile, the native habitat of monkey-puzzle (*Araucaria araucana*), which is also often seen in the landscape around Florida.

Size

What was once a large family is now reduced to just three genera: *Agathis*, *Araucaria*, and *Wollemia*. Cool-climate gardeners will be most familiar with the monkey-puzzle (*Araucaria araucana*), which has been widely introduced to yards worldwide because of its hardiness and unique structure. Warm-climate gardeners may be better acquainted with the Norfolk Island pine (*Araucaria heterophylla*), where along coastal areas it is widely planted as a street tree.

Origins

This family was at its peak during the Carboniferous (more than 300 million years ago) and was at its most diverse during the Jurassic and Cretaceous (between 200 and 66 million years ago). Fossil records suggest the extinction of the dinosaurs and the collapse of *Araucariaceae* coincide.

Araucaria araucana,
monkey-puzzle tree

Leaves

The evergreen leaves are arranged in spirals
around the stem, and they persist for years.
Many species have small, curved, needlelike leaves
that tend to be soft to the touch, but the broad
and thick, sharply pointed scales of the monkey-
puzzle are nothing but hostile. The broad, leathery
leaves of *Agathis* trees do not look like conifer
leaves at all.

Cones

The large, round female cones are of particular
interest because of their huge size. Borne on
upright stems at the tips of branches, they look
unusual and can cause serious damage to anyone
wandering beneath the tree when they fall. The
seeds of some species are considered a delicacy.

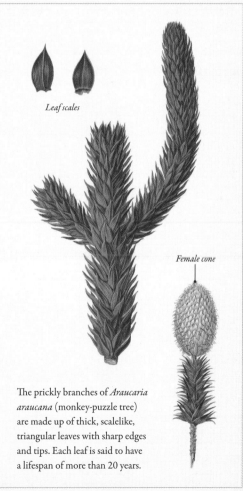

Leaf scales

Female cone

The prickly branches of *Araucaria
araucana* (monkey-puzzle tree)
are made up of thick, scalelike,
triangular leaves with sharp edges
and tips. Each leaf is said to have
a lifespan of more than 20 years.

USES FOR THIS FAMILY

Commonly planted as landscape trees, plants in
this family have a strong architectural outline
but are rather too big for anything other than
large gardens. Commonly grown species include
the monkey-puzzle (*Araucaria araucana*), Norfolk
Island pine (*A. heterophylla*), bunya-bunya pine
(*A. bidwillii*), and hoop pine (*A. cunninghamii*).
The Wollemi pine (*Wollemia nobilis*) is being
trialed in cultivation, and one day could become
as widespread and popular as the others.

Podocarpaceae
THE PODOCARP FAMILY

Until recently, many authorities questioned whether the podocarp family were indeed true conifers. Anatomical anomalies, such as their fleshy "berries," had previously suggested they belonged to a separate grouping altogether. However, DNA evidence now places them with the true conifers.

Size

While there are 172 species described, gardeners will be most familiar with the genera *Dacrydium* (rimu), *Phyllocladus* (celery pine), *Podocarpus*, and *Prumnopitys*. All are evergreen shrubs and trees.

Range

The *Podocarpaceae* is largely a southern hemisphere family, centered in the Australasian territories of New Zealand, New Caledonia, Tasmania, and South America.

Origins

Vast forests of podocarps once thrived in Antarctica, back when it formed part of the supercontinent Gondwana with a climate much warmer and wetter than today's. Gondwana began to break up during the early Cretaceous, leaving isolated pockets of these prehistoric survivors dotted about the southern hemisphere.

Leaves

The individual leaves of podocarps are arranged in spirals, either in two ranks or as tiny scales. The genus *Phyllocladus* (celery pines) are a little different because the tiny scales soon turn brown and drop off. In this case, the highly modified, leaflike shoots, called phylloclades, are left to do the job of photosynthesis, giving the plants their characteristic ferny appearance—rather like the fronds of celery.

Phyllocladus hypophyllus,
Malaysian celery pine
While the ferny, celery-like phylloclades of these trees are unusual and distinctive, they are more keenly appreciated for their timber.

Prumnopitys montana,
diablo fuerte or strong devil

USES FOR THIS FAMILY

In yards, podocarps can be grown as specimen trees or clipped as hedges. These are popular in the Southeastern United States and in California. The fruits have a sweet flavor, but are still toxic, so food preparations made with them should only be eaten in small amounts.

Poisonous properties

Like their close relatives, the yew family, all parts of podocarps are poisonous. Heavy exposure to podocarp pollen, released in spring and early summer, can produce symptoms that mimic the toxic side effects of chemotherapy. Contact with the male plants can also trigger allergies.

Fruit

It is easy to lump together the podocarp and yew families (*Taxaceae*—see pages 64–65), since the female plants both bear the same fleshy "berries" (though they are not berries in the strict botanical sense). In fact, the name "podocarp" is derived from the Greek *podus* (foot) and *karpos* (fruit), in reference to the berrylike fleshy fruits and stalks that feature on many species. The fruit of each genus is quite distinctive, which is a useful aid to identification. Sometimes the seed is completely enclosed by the fleshy aril (*Prumnopitys*), and at other times the seed sits atop the aril (*Podocarpus nivalis*). Occasionally the fruit is accompanied by a fleshy, colorful receptacle or stalk (*P. dispermus*).

Podocarpus neriifolius,
brown pine
The curious fruits of these plants have the appearance of berries. The seed has a fleshy covering and sits atop a fleshy receptacle.

Pinaceae
THE PINE FAMILY

From the stately cedars to the classic Christmas tree, the beautiful landscape trees of the pine family are familiar to all by their cones and needles. They are also the largest family of conifers. Almost all of them are evergreen, with two notable exceptions: *Larix* (larch) and *Pseudolarix* (golden larch).

Size

Approximately 225 species make up the pine family. This can be broken down into 12 genera, seven of which are immediately familiar: firs (*Abies*), cedars (*Cedrus*), larches (*Larix*), spruces (*Picea*), pines (*Pinus*), Douglas firs (*Pseudotsuga*), and hemlocks (*Tsuga*).

Picea abies,
Norway spruce

Range

In geographical terms, the pine family ranges across the whole of the northern hemisphere, with most species in temperate regions. Subarctic outliers include the Scots pine (*Pinus sylvestris*), which is the only pine native to northern Europe, and covers vast forested areas from Europe to eastern Siberia, well into the Arctic Circle, often forming stands with its family member, the Norway spruce (*Picea abies*). The Jack pine (*Pinus banksiana*), a native of Canada, is the most northerly occurring species. There are many subtropical *Pinaceae*, especially in Mexico, but only one species crosses the equator: *Pinus merkussi* on Sumatra. While you are likely to find a representative of the *Pinaceae* growing close to wherever you are in the temperate world—maybe even in your own yard—the main centers of biological diversity are in North America (including Mexico), China, and Japan.

Cedrus atlantica,
Atlas cedar

Origins

While conifers first appear in the fossil record approximately 300 million years ago, fossils taken from the Jurassic show that the first recognizable *Pinaceae* began to appear about 200–150 million years ago, when conifers were the dominant land plants and a key food source for the dinosaurs. The species and genera we see today evolved from these ancestral pines, as they colonized new territory and adapted to new environments.

Leaves

Pinaceae all have needlelike leaves, borne singly or in small bundles, clusters, or tufts. These are very distinctive, and the only plant that might be mistakenly included in this family is the Japanese umbrella pine (*Sciadopitys verticillata*), an anomalous species that belongs to its own family. With its needles arranged in umbrella-like whorls, the similarity is obvious, but umbrella pines have their needles clustered at the branch tips, a trait not seen in the pine family.

Fig.1.1

Fig.1.2

Fig.1.1. Larch species (e.g. *Larix decidua*) always have their needles in tufts.

Fig.1.2. True pine (*Pinus banksiana*) have their needles in small bundles.

Pinus sylvestris, Scots pine

The arrangement and characteristics of the needles is an important way to tell the species apart. Cedar and larch species always have their needles in tufts and clusters, whereas other species, such as the true pine (*Pinus*) have their needles in small bundles (see figs 1.1 and 1.2, above). Of the single-needled species (*Picea, Abies, Tsuga, Pseudotsuga*), the *Tsuga* are distinct, with their needles arranged in two ranks.

Trees from the spruce (*Picea*), fir (*Abies*), and Douglas fir (*Pseudotsuga*) genera make the most popular Christmas trees. Spruces and firs are easily told apart by stroking their foliage; if it is soft and thick it is probably a fir tree, and if it is sharp and prickly, then it must be a spruce.

Picea glauca,
white spruce

Female cone

The difference between spruces and firs is seen in the cones: fir cones stand upright on the branches and spruce cones hang down. People familiar with these trees will also note that the needles of firs are much softer and less prickly compared to the needles of spruces, which are more slender.

Cones

Everyone knows what a pine cone looks like, but it is the way in which the cone scales overlap each other, like fish scales, that sets them apart from the cones of other conifers. The botanical word for this is "imbricate." Pine cone shapes vary from that of an egg, to elongated, conical, and almost cylindrical. The size ranges from ¾ to 15 ¾ inches in length; from the small, grape-size cones of the larch tree, to the giant pineapple-size cones of the Coulter pine (*Pinus coulteri*)—people are actually advised to wear hard hats in Coulter pine groves.

In some genera, the cones are held upright on the branches (*Abies, Cedrus*), and in others they droop (*Picea, Pseudotsuga*). This can help with identification: Douglas firs may look like spruces, but what gives them away is the unusual, three-lobed "mouse tail" bracts that stick out from between the cone scales.

Pinus muricata,
Bishop pine

Abies balsamea,
balsam fir

Pinus coulteri,
Coulter pine

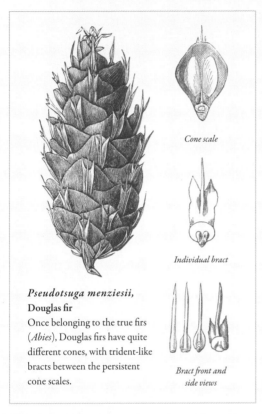

Cone scale

Individual bract

Pseudotsuga menziesii,
Douglas fir
Once belonging to the true firs
(*Abies*), Douglas firs have quite
different cones, with trident-like
bracts between the persistent
cone scales.

*Bract front and
side views*

USES FOR THIS FAMILY

From pine nuts to turpentine, Christmas trees to
rosin for violins—not to mention their timber
value—without the pine family, human life
would be very much the poorer. As ornamental
landscape trees this family excels, with good
evergreen foliage and a well-defined crown
shape. The large number of species means that
there will be at least one that suits your climate
and soil condition, and many tolerate drought.
Their only demand is full sun. Pines also provide
the best trees for coastal shelter, and for small
yards there are many dwarf cultivars, such as
Pinus mugo "Mops" and *Abies balsamea* "Nana."

Pseudotsuga menziesii,
Douglas fir

The cones take anywhere from six months to
two years to ripen. When they finally ripen, the
seeds are released either when the cone scales
flex back as they dry out or when the cones
themselves disintegrate (*Abies* and *Cedrus*).
In some species, the cone scales will only open
when exposed to extreme heat or fire. This is
an environmental adaptation to fire-prone
environments. The three "closed-cone pines"
of California—Bishop pine (*Pinus muricata*),
Monterey pine (*P. radiata*), and knobcone pine
(*P. attenuata*)—are good examples, but it is also
seen in other regions, such as the Mediterranean,
home of the Aleppo pine (*P. halepensis*). After a
forest fire, the seeds are released, ready to
germinate in the first rain and colonize the
newly scorched landscape.

Cupressaceae
THE CYPRESS FAMILY

The cypress family is made up of the "frondy" conifers, sometimes mistakenly referred to as fir trees, but actually quite different from true firs (*Abies*), which belong to the pine family (*Pinaceae*). It is a family of evergreen trees and shrubs, including the infamous Leyland cypress (× *Cuprocyparis leylandii*).

Size

This is a relatively large conifer family of approximately 130 species, some of which are quite obscure, though others are very widely grown in temperate regions around the world.

Of the 30 genera, the best-known are the false cypresses (*Chamaecyparis*), true cypresses (*Cupressus*), junipers (*Juniperus*), dawn redwood (*Metasequoia*), giant sequoia (*Sequoiadendron*), coast redwood (*Sequoia*), and arbor-vitae (*Thuja*). With the exception of *Metasequoia*, all of these are evergreen trees. In cultivation there is a wide range of shrubby and dwarf forms, as well as creeping and ground-covering shrubs.

Range

There are not many habitats that have not been colonized by the *Cupressaceae*, with the exception of tropical rain forest and arctic tundra. Their geographical range stretches from the Arctic Circle in Scandinavia to the southernmost tip of South America. In fact, *Pilgerodendron uviferum*—native of Tierra del Fuego—is the world's southernmost conifer; and black juniper (*Juniperus indica*)—found in the northern Himalayas—is the highest-altitude tree in the world.

Common juniper (*Juniperus communis*) possibly has a wider distribution than any other shrub, being common throughout Europe, Asia, and North America, while the coast redwood (*Sequoia sempervirens*) is restricted to a narrow belt of humid coast territory about 450 miles long and 20 miles wide, extending from the southwestern corner of Oregon to Monterey County in southern California, where it is the dominant tree among other conifers. The dawn redwood (*Metasequoia glyptostroboides*) only existed as a fossil until 1941, when living plants were discovered in a very restricted habitat in Hubei Province, China. Like a number of other species in *Cupressaceae*, dawn redwood is endangered in its native habitat.

***Cupressus sempervirens*,**
Italian cypress

Juniperus sabina,
savin

Juniperus communis,
common juniper

treasures during the 18th and 19th centuries, for the first time in millions of years distant family members were brought together in the collections of botanic gardens. Some interesting hybrids arose, such as between Japanese and North American arbor-vitaes (*Thuja*) and the fast-growing, sometimes notorious, Leyland cypress (× *Cuprocyparis leylandii*). The latter is a cross that was first noticed when *Chamaecyparis nootkatensis*—a native of the Pacific Northwest—and the Monterey cypress (*Cupressus macrocarpa*)—a native of California—were grown in close proximity for the first time at Leighton Hall in Wales, UK.

Sequoiadendron giganteum,
giant redwood
The sole living species in their genus, these are the world's largest trees, confined to a small area in California's Sierra Nevada.

Origins

Fossils from the Triassic and Jurassic show how conifer cones gradually evolved over millions of years. A seed cone fossil from the Isle of Skye, western Scotland, provides the oldest known evidence for the *Cupressaceae*, dating the emergence of this family to about 170 million years ago.

As the continental plates drifted apart, the climate changed and the landscape reformed itself through earthquakes and volcanic activity, *Cupressaceae* family members became separated from each other, following their own evolutionary paths in isolation. When plant hunters began exploring the continents for botanical

Leaves

Most *Cupressaceae* are easy to recognize by their frond-like foliage, which lends them a slightly shaggy appearance. They are made up of tiny, scalelike leaves, arranged in opposite groups—usually opposite pairs but sometimes in threes or fours. Notable exceptions include *Metasequoia* and *Sequoia*, which have longer needles arranged in two ranks, a characteristic that has, in the past, led some botanists to place them in the yew family (*Taxaceae*—see pages 64–65).

Mature seed cone

Cone scale with seeds

Shoot with buds

Sequoia sempervirens, coast redwood

Pollen-producing scale

Cone after seed release

Male cones

Chamaecyparis lawsoniana,
Lawson's cypress
The globe-shaped cones of *Chamaecyparis* are small, the size of a pea.

USES FOR THIS FAMILY

In terms of horticulture, the cypress family, in particular the junipers, false and true cypresses, and arbor-vitaes are hugely important, with hundreds of cultivated varieties available, from the shrubby ground covers to the upright Italian cypresses and other, taller landscape trees. The bark can be quite distinctive, commonly orangey brown, peeling in strips with a stringy texture. Many species also clip well, making dense hedges and screens, shapely shrubs, or even topiary: for example, *Chamaecyparis lawsoniana*, x *Cuprocyparis leylandii*, *Cupressus macrocarpa*, *Thuja plicata*, and *T. occidentalis*.

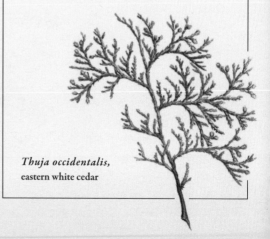

Thuja occidentalis,
eastern white cedar

While it can be difficult to tell whether opposite pairs of the scale leaves are in twos or fours (sometimes they are closely compacted together), junipers are easy to spot, with their scales arranged in sets of three. This, and their attractive berries, makes them easy to identify. Many *Cupressaceae* have needlelike leaves when young, which turn to scales on maturity. Some species of juniper (*Juniperus communis*) retain needlelike leaves into maturity.

Cones

Juniperus is an anomaly among the *Cupressaceae* because it bears berrylike fruit; all the other species bear cones. For the rest, it is the anatomy of their cones that gives these trees away as belonging to the cypress family (as opposed to the pine family, for example). Perhaps the best way to get to know the individual members of this family is to study their small cones. These turn from green to brown and woody as they mature, changing shape as the scales open to release the seeds within.

A few of the lesser-known species, such as the incense cedars (*Calocedrus*), have cones that are segmented like an orange, and these spread open to a star shape. Globe-shaped cones belong to either the true or false cypresses (*Cupressus* or *Chamaecyparis*), with those of the false cypresses being pea-size, and those of the true cypresses much bigger. *Thuja* cones are also small, like *Chamaecyparis*, but become bell-shaped when open.

Metasequoia, *Sequoia* and *Sequoiadendron* cones are egg-shaped, more like a cross between pine cones and cypress cones. The scales do not overlap—as they do with the *Pinaceae*—and they typically remain green and closed for many years.

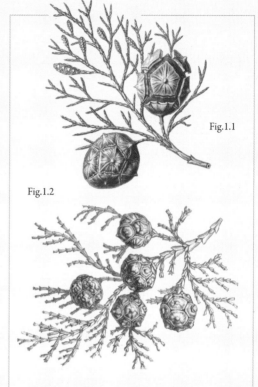

Fig.1.1

Fig.1.2

Fig.1.1. *Cupressus sempervirens* (Mediterranean cypress) is a true cypress, with rounded seed cones 1–1½ inches in diameter and 10–14 scales.

Fig.1.2. *Chamaecyparis obtusa* (Japanese cypress) is a false cypress, with rounded seed cones about ½ inch in diameter and 8–12 scales.

Fig.1.3. The huge seed cones of *Sequoiadendron giganteum* (giant redwood) are 1½–2¾ inches in diameter with 30–50 spirally arranged scales.

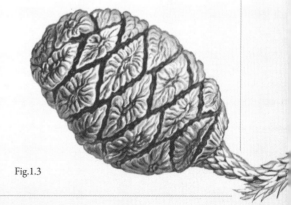

Fig.1.3

Taxaceae
THE YEW FAMILY

The single distinctive feature of the yew family is a lack of cones. Rather, plants in this family bear single ovules or seeds, enclosed by a fleshy aril. This led botanists to separate them from the true conifers, but DNA analysis now shows that they are more closely related than previously thought.

Size

This relatively small family contains 28 species of small- to medium-size trees and shrubs. Nutmeg yew (*Torreya*), true yew (*Taxus*), and plum yew (*Cephalotaxus*) are three of the better-known genera.

Range

The yew family is stretched out right across the northern hemisphere, across North America and Asia, and down into southeastern Asia and Central America. The species *Austrotaxus spicata*—an outlier in the Pacific Island of New Caledonia—shows close family links to the podocarp family (*Podocarpaceae*—see pages 54–55).

Origin

Plants of the *Taxaceae* are all essentially quite similar, descended from a common ancestor, *Paleotaxus* (now extinct), then separated by geographical boundaries. Many millions of years of evolution led to diversification into several different genera, some of which survive to this day.

Taxus baccata "Fastigiata," Irish yew

USES FOR THIS FAMILY

The yew has immense value in the yard as a clipped shrub, screen, or hedge. Its moderate growth, deep color, and neat habit make for a fine plant. As a tree, it is more unruly, casting deep shade, and is often found in churchyards. The plant is highly poisonous, and is used in the anticancer drug taxol.

Seed

Seed in section

Torreya,
nutmeg yew

Taxus baccata,
English yew

Cephalotaxus,
plum yew

Leaves

There is a high degree of uniformity among the family, which is to say they all look very much the same to the untrained eye. The dark, evergreen leaves are arranged in spirals along the stem, but most often twisted at their bases, so that they appear to be in two ranks. This leaf arrangement can cause confusion with other conifers—most notably the sequoias, sequoiadendrons, and metasequoias of the cypress family—which bear cones and not seeds with fleshy arils. Nutmeg yews (*Torreya*) have sharply pointed leaves; true yews (*Taxus*) do not.

Fruit

The fruits are colorful, and when set against the dark, sometimes somber foliage, they can be a decorative feature in their own right. Only the female trees produce fruits, which contain one seed surrounded by a fleshy coat (aril). The colorful aril partially, or almost completely, encloses the highly poisonous seed. Birds are attracted to the bright arils, which are sweet and juicy; the seed passes through their gut undamaged, and is thus dispersed in their droppings.

   Typically among yews the arils are red. Exceptions include the white berry yew (*Pseudotaxus*), which has white arils. Nutmeg yews have seeds with thin arils, appearing like olives. The seeds of the Japanese *Torreya nucifera* are actually edible, and considered a local delicacy.

Lilium candidum,
Madonna lily
From the lily family,
(*Liliaceae*—see pages 78–79).

CHAPTER 2

Monocots and Early Angiosperms

Much of the decorative appeal of garden plants results from their beautiful blooms. Angiosperms are the only plants to produce flowers, and while animal pollinators are the target audience, gardeners have learned to appreciate their charms.

Angiosperms appeared around 130 million years ago, and descendants of some of those early pioneers survive today, such as the waterlilies and magnolias. These early offshoots have a lot to offer gardeners because they often have large flowers with many colorful petals and a variety of growth habits.

The angiosperms are by far the largest group of plants on Earth today. They are readily divided into two major groups: monocots and eudicots. Despite being the smaller of the two, monocots pack a punch. Not only do they include crucial crops such as grains, but also their impact in the garden is impressive. Ornamental grasses; exotic orchids; fragrant lilies; many spring bulbs; and the exciting foliage of hostas, palms, and bananas are all part of this diverse assemblage. Look out for flowers with parts in threes and leaves with parallel veins, which are both identifying features.

Nymphaeaceae
THE WATERLILY FAMILY

These perennial aquatic plants are easily recognized, thanks in part to the French Impressionist Claude Monet. They are also widely appreciated for the sense of tranquillity they bring to ponds and lakes.

Size

Around 95 species are known, and most reside in the widely cultivated genus *Nymphaea*. Also familiar are the pond lilies (*Nuphar*) and giant waterlilies, the fox nut (*Euryale*), and the Amazon waterlily (*Victoria*).

Range

Most early angiosperm families are relicts, restricted to isolated islands, but *Nymphaeaceae* have spread across the globe. They're found on all continents except Antarctica, though they are generally absent from polar, tundra, and desert regions.

Origins

Fossilized waterlilies date back 90 million years to the Late Cretaceous. It should be noted that the sacred lotus (*Nelumbo nucifera*), while superficially similar to waterlilies, has more recent origins, and is placed in its own family, *Nelumbonaceae*.

Flowers

The characteristic blooms of waterlilies are solitary, and either float on the surface, or rise above it. They have six to twelve sepals—some green, some colorful—and eight or more petals. The petals intergrade with the numerous stamens, and some petals have partially formed anthers. A large, disc-shaped stigma sits in the center of the flower.

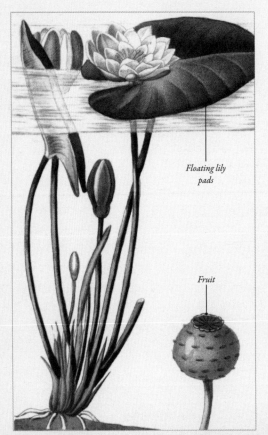

Floating lily pads

Fruit

Nymphaea alba,
European white waterlily

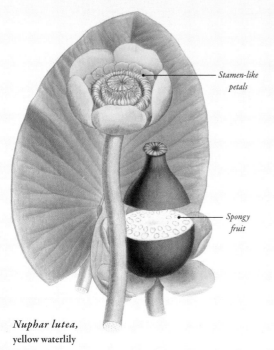

Stamen-like petals

Spongy fruit

Nuphar lutea,
yellow waterlily

USES FOR THIS FAMILY

No pond is complete without a waterlily, but they prefer still water, where their leaves will not be inundated, so turbulence from streams or fountains should be avoided. It is also important to choose your lily carefully because larger species, such as the European white (*Nymphaea alba*), can readily overwhelm a smaller pond. When buying, look for a label describing the water depth for that species and choose one that corresponds with the depth of your pond.

Fruit

The spongy fruits swell up and split open, releasing the seeds into the water. The bottle-shaped fruits of *Nuphar* remain buoyant, while in other waterlilies the fruits sink below the water surface.

Leaves

The floating foliage is typically round, though it can be narrow or arrow-headed. *Nuphar* species often have numerous lettuce-like, submerged leaves. Hairs on the surface produce mucilage, a viscous substance, giving the pads a slimy texture. The leaves emerge from a large horizontal stem— or rhizome—growing on the lake bed. The massive leaves of *Victoria* and *Euryale* are supported by a network of ribs, and the lower leaf surfaces are covered in spines. *Victoria* leaves also have upturned edges, and the massive pads are said to be strong enough to support the weight of a small child.

Habitat

As the name suggests, most waterlilies live partly or fully submerged, but the dwarf waterlily (*Nymphaea thermarum*), the world's smallest waterlily, hails from wet mud on the edge of a hot spring in Rwanda. Unfortunately, local farmers diverted the water for agriculture, destroying its habitat, so this plant now only exists in botanic gardens.

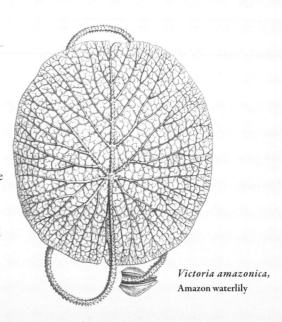

Victoria amazonica,
Amazon waterlily

Magnoliaceae
THE MAGNOLIA FAMILY

With their often large, luxurious blooms, this distinctive family of trees and shrubs is not easily confused with any other. While it includes several large trees, more compact species are available that will fit almost any garden.

Size

Historically, the 221 species in this family were divided among 12 or more genera, but today, only two are recognized. There are two species of tulip trees (*Liriodendron*), with the remainder categorized as *Magnolia*.

Range

Magnolias are only found in southeastern Asia and the Americas, and can occur in both tropical and temperate regions. This pattern is mirrored by the tulip trees, with one species (*Liriodendron tulipifera*) in the eastern regions of North America and the other (*Liriodendron chinense*) in China.

Origins

A handful of fossils date *Magnoliaceae* back 105 million years to the Cretaceous, though fossils increase in abundance from the later part of this period. Seed fossils attributable to *Liriodendron* first appear around 93.5 million years ago, and fossils of both genera can be found in Europe, outside their modern ranges.

Flowers

The spectacular flowers of *Magnoliaceae* are so striking in part because they appear before the leaves do in spring. Of course, this is not the case in all species, especially the evergreens. The blooms are solitary, and initially enclosed in one or more papery, often hairy bracts, which fall off as the bud expands. Each flower has a cone-shaped receptacle onto which all floral parts are spirally arranged.

Magnolia denudata,
lily tree

Fruit

Once pollinated, the carpels swell up, often merging with their neighbors to produce a peculiar aggregate fruit. Mature carpels are known as follicles, and each splits open to release a single seed with a red or orange fleshy coat. They dangle on a thread of tissue and are snapped up by hungry birds. Tulip trees have dry, winged seeds that are distributed by wind.

Leaves

Compared to the flowers, magnolia leaves are often rather unspectacular, being alternate and simple. Evergreen leaves may have waxy or hairy undersides and stipules can be found with younger leaves, though these readily fall off. Tulip trees, on the other hand, have very distinctive leaves with four or six lobes and flat or notched tips.

USES FOR THIS FAMILY

Where unlimited space is available, grand trees such as deciduous Campbell's magnolia (*Magnolia campbellii*), southern magnolia (*M. grandiflora*), and either of the tulip trees make for great features. Smaller species, such as the dainty *M. stellata*, or *M. sieboldii*, with its downward-hanging blooms, work well in more confined yards, while cultivars of *M. laevifolia* and banana shrub (*M. figo*) can be pruned to form fragrant hedges.

Magnolia campbellii,
Campbell's magnolia

All magnolia flowers, such as the one pictured here on *Magnolia campbellii*, have a cone-shaped receptacle in the center, to which the petals, stamens, and carpels are attached in spiral formation.

The sepals and petals are usually indistinguishable, numerous, and colorful. Stamens are also plentiful with short, stubby filaments. At the center of the bloom is a cluster of carpels, each with a hooklike stigma.

Liriodendron tulipifera,
American tulip tree

Araceae

THE ARUM FAMILY

Members of the arum family are collectively known as aroids, and they have one major identifying characteristic: their instantly recognizable flowers. A few species are grown as houseplants. Many are poisonous. Typical growth is herbaceous (nonwoody) with thick underground tubers or rhizomes.

Size

The *Araceae* is a large family of about 3,250 species within 105 genera. They range widely in appearance and size, and for this reason some authorities further organize the family into a number of subfamilies.

Philodendron verrucosum, velvet-leaf philodendron

Range

Predominantly a tropical family, with a worldwide distribution, there are a few notable temperate species, such as cobra lilies (*Arisaema*), lords-and-ladies (*Arum maculatum*), mousetail plant (*Arisarum proboscideum*), dragon arum (*Dracunculus vulgaris*), and arum lily (*Zantedeschia aethiopica*).

Origin

The *Araceae* has one of the oldest fossil records among the flowering plants, with species beginning to diversify from the Early Cretaceous (140–130 million years ago). Watery habitats appear to be ancestral in the family, which explains why a large number of aroids tolerate wet or very moist soil. *Zantedeschia aethiopica* and *Lysichiton americanus*, for example, both make very fine marginal aquatics for the garden.

Invasive species

Water lettuce (*Pistia stratiotes*) is a notable weed in many countries, because its vigorous growth can clog waterways and destroy natural habitats. Gardeners should avoid this plant and dispose of any existing plants very carefully.

Arisaema utile,
cobra lily

Leaves

The leaves are simple, lobed, or divided, and
emerge from the base of the plant or on aerial
stems. Dumb cane (*Dieffenbachia*), Swiss cheese
plant (*Monstera deliciosa*), and Chinese ivy
(*Philodendron*), among others, make good foliage
houseplants. The leaves of the Swiss cheese plant
develop large holes as they grow.

USES FOR THIS FAMILY

The various forms of taro plant—elephant's ear
(*Alocasia*), cocoa root (*Colocasia*), and yautia
(*Xanthosoma*)—are important staple food crops
in the tropics, grown for their starchy tubers.
Many are grown as ornamental plants, both
indoors and out, and the cut flowers of banner
plant (*Anthurium*) and arum lily (*Zantedeschia*)
are prized by florists.

Flowers

The shape of *Araceae* inflorescence is
unmistakable, sometimes hooded like a cobra
(*Arisaema*), dark and somber like a dragon, or
bright and colorful like a banner (*Anthurium*).
Those of the mousetail plant (*Arisarum
proboscideum*) have long and narrow tips, like a
mouse's tail. The titan arum (*Amorphophallus
titanum*) from southeastern Asia has the largest
single inflorescence in the world.

Araceae inflorescences are made up of a
flowering spike called a spadix, surrounded
by a petal-like bract called a spathe. The spadix
of many species emits a repulsive odor that
attracts flies and beetles for pollination.
Sometimes the spadix heats up, so the
perfume is spread more widely.

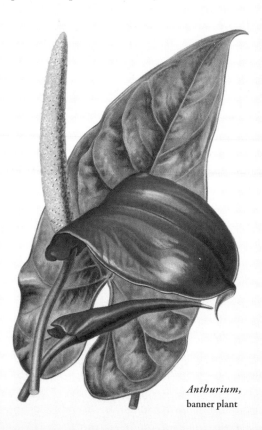

Anthurium,
banner plant

Melanthiaceae

THE WAKE-ROBIN FAMILY

Until quite recently, plants of the *Melanthiaceae* were part of the lily family (*Liliaceae*), but new studies have found there to be significant hereditary differences in their DNA. Since the plants, flowers, and cultivation requirements of both families are very similar, gardeners commonly treat them as one group.

Size

Compared to the lily family, the *Melanthiaceae* is relatively modest with just 120 species of herbaceous plants, typically producing flowers on upright spikes. The evocative common names of the 16 different genera give a clue to the nature of this family: swamp pink (*Helonias bullata*), blazing star (*Chamaelirium luteum*), feathershank (*Schoenocaulon*), feather bells (*Stenanthium*), false hellebore (*Veratrum*), bear grass or turkey beard (*Xerophyllum asphodeloides*), and wake-robin (*Trillium*).

Xerophyllum asphodeloides,
turkey beard

The family name derives from the genus *Melanthium*, which is composed of just four species of bulbous plant known as bunchflowers. The best-known family member, however, must be *Trillium*, with its pretty flowering plants found on woodland floors. Similar but less showy is the closely related genus *Paris*. Another popular group is *Veratrum*, often found in wet mountain meadows.

Range

North America seems to be the ancestral home of this family, but its range extends across Asia, with a few European representatives. A handful of American species can be found in Central America, and a couple in the northern regions of South America. Of all the genera, *Trillium* is probably most widespread.

Helonias bullata,
swamp pink

Origin

It is clear that the *Melanthiaceae* shares a common ancestry with the lily family (*Liliaceae*), which first appeared in the fossil record around about the time of the Cretaceous–Palaeogene extinction. Evidence suggests that the *Melanthiaceae* began to follow its own course of evolution fairly early on in the lily family lineage.

Roots

Melanthiaceae are herbaceous plants, and a large percentage have swollen storage organs, such as bulbs, corms, rhizomes, or thick and fleshy roots. In temperate areas, the above-ground parts will die back during harsh weather in summer or winter (depending on the habitat), and the plant will persist underground until the conditions become more favorable.

Flowers

With the exception of herb paris (*Paris quadrifolia*), plant parts of the *Melanthiaceae* are arranged in multiples of three. Thus, we have the consistently six-petaled flowers of *Veratrum*, and the very uniform three-parted flowers of *Trillium*. Each upright *Trillium* stem bears three opposite-facing leaf bracts, in the center of which sits the flower with three sepals, three petals, two pairs of three stamens, and three stigmas. Herb Paris is similar, but its parts are in fours.

Paris quadrifolia,
herb Paris
With its leaves and flower parts in fours, herb paris is distinguishable from *Trillium*, with which it is sometimes confused.

USES FOR THIS FAMILY

Many species and cultivars of *Trillium* make good woodland garden plants; *Veratrum* is better in sun, as long as the soil is reliably moist. Many species are poisonous; Native Americans knew this and used root extracts as insecticides, medicines, and arrow poisons.

Veratrum nigrum,
black false hellebore

Trillium grandiflorum,
American wake-robin

Colchicaceae
THE AUTUMN CROCUS FAMILY

Like *Melanthiaceae*, the autumn crocus family is a recent casualty of the botanical review of the lily family (*Liliaceae*—see pages 78–79), when 15 genera were separated out and the *Colchicaceae* became a formally recognized family. *Colchicaceae* is composed entirely of small to medium flowering plants that grow from underground tubers, rhizomes, or corms.

Size

Of the 225 species described, about two-thirds belong to the genus *Colchicum*. The lesser-known genus, *Wurmbea*, contains about 50 species.

Range

While the majority of this family seems to be centered around southern Africa, *Colchicum* (meadow saffron) is distributed quite widely across the African continent, into Europe and western Asia, spreading up through Africa and into the Mediterranean. *Disporum* (fairy bells) has representatives from northern India to Japan and southeastern Asia. There are also a few Australian natives. *Uvularia* (bellwort) is the sole American member, with four species from the United States and eastern Canada.

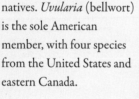

Disporum cantoniense,
fairy bells

Origin

Like the *Melanthiaceae*, the *Colchicaceae* shares a common ancestry with the lily family, with the two families diverging from the lilies and following their own evolutionary path quite early on in the process of evolution, about 60 million years ago. Another family that followed this path is the Peruvian lily family (*Alstroemeriaceae*).

Flowers

The late summer and fall flowers of *Colchicum* are often mistaken for spring crocuses, whose timing has gone a bit awry. True crocuses (*Crocus*) belong to the iris family (*Iridaceae*—pages 82–83), and gardeners can tell them apart by counting the stamens: *Iridaceae* members usually have just three, whereas *Colchicaceae* species most often have six. Confusingly, there are also a number of fall-flowering *Crocus*, such as large autumn crocus (*Crocus speciosus*).

Others are almost lily-like, such as the various species of *Disporum*, and the Chinese lantern lily (*Sandersonia aurantiaca*) with its bright orange flowers.

The latin name *Uvularia* possibly derives from the flowers—or merrybells—which dangle in a way that is reminiscent of the uvula—the soft organ that hangs down at the back of the throat.

Colchicum autumnale,
autumn crocus

Leaves

The leaves of most of the *Colchicaceae* species tend to be much longer than they are wide, like the foliage of many garden bulbs. *Disporum* species have broader, elliptic leaves. All have parallel veining. In the *Gloriosa* (glory lily), tendrils are produced at the leaf tips to help the plants scramble and climb.

To survive through the hot African summers, or harsh Asian or European winters, the above-ground parts of many species will die back. Typically, the leaves will emerge again when the seasons become favorable, to be joined soon after by the flowers. *Colchicum* leaves die back before the flowers arrive in late summer or fall, giving the blooms an unclothed appearance. Some call them "naked ladies."

Gloriosa superba,
glory lily

Sandersonia aurantiaca,
Chinese lantern lily

Liliaceae

THE LILY FAMILY

Spring just wouldn't be the same without the many bulbous plants in *Liliaceae*, which include tulips (*Tulipa*), fritillaries (*Fritillaria*), and trout lilies (*Erythronium*). The true lilies (*Lilium*) extend the colorful display into summer and fall.

Tulipa gesneriana,
common garden tulip

Size

The revolution in DNA research wrought huge changes in many plant families, and none more so than *Liliaceae*. At its peak, it included around 4,500 species in 300 genera, but now is reduced to a mere 600 species in 15 genera.

Range

Restricted to the northern hemisphere, *Liliaceae* shows greatest diversity in the Himalayas and China. Plants occur in grassland, woodland, and alpine meadows. They are all herbaceous perennials, derived from bulbs or rhizomes.

Origins

Liliaceae fossils are said to date back to the Late Cretaceous (100–70 million years ago), though the identification of these fossils is debatable, especially considering the great taxonomic changes in this family.

Flowers

The often extravagant blooms of this family each have three sepals and three petals, which are usually similar and often decorated with stripes or spots. In *Scoliopus,* however, there are only three petals, while in mariposa lilies (*Calochortus*) the sepals and petals are distinct. In toad lilies (*Tricyrtis*)

Petal crests

Nectar
pouches

*Tricyrtis
macropoda,*
**large-stalked
toad lily**

Clintonia borealis,
bluebead

USES FOR THIS FAMILY

Liliaceae is a popular family in the garden, with many uses. Tulips can be planted in containers, beds, and borders for spring color and then lilies can be used in the same way for a colorful display through the summer. The fall blooms of toad lilies complete the year. In woodland gardens, the giant lily (*Cardiocrinum*) makes a bold statement, while smaller bluebead lilies (*Clintonia borealis*), *Nomocharis, Prosartes,* and *Streptopus* reward closer inspection. And where a pristine lawn is not required, species tulips (*Tulipa sylvestris, T. sprengeri*) and snake's head fritillaries (*Fritillaria meleagris*) can be naturalized.

the three sepals have nectar-secreting pouches at the base while the petals have prominent upright crests. Six protruding stamens, which sometimes produce copious amounts of pollen, are the norm—although *Scoliopus* has only three. Some species have scented flowers, though these can be either sweet (some *Lilium*) or downright unpleasant (crown imperial, *Fritillaria imperialis*).

Fruit

The fruits of *Liliaceae* are usually capsules that split open when dry. Blue, purple, or black berries are produced by bluebead (*Clintonia*) and cucumber root (*Medeola virginica*).

Leaves

The simple leaves of *Liliaceae* are typically arranged in a rosette at the base of the plant, or alternate along the unbranched stem. In a few species—Turk's cap (*Lilium martagon*), black sarana (*Fritillaria camschatcensis*), and cucumber root (*Medeola virginica*)—the leaves are whorled. Most leaves have little or no discernible petiole.

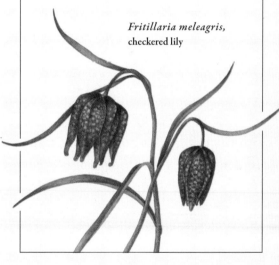

Fritillaria meleagris,
checkered lily

Pests

A great threat to garden lilies is the scarlet lily beetle (*Lilioceris lilii*), native to parts of Europe and Asia and now widespread across the northern hemisphere. These red terrors and their larvae devour lilies and fritillaries, causing considerable and unsightly damage. Adults and larvae should be picked off whenever they are found, often lurking beneath the foliage.

Orchidaceae

THE ORCHID FAMILY

Prized for their spectacularly beautiful flowers, orchids are cultivated with sometimes fanatical devotion. Although there exists an enormous number of species and hybrids, all orchids can be classified as either living on the ground (terrestrial) or in the branches of trees (epiphytes).

Size

The orchid family is very large, containing more than 18,000 species. Despite this, it shows much less diversity than many smaller families.

Spathoglottis plicata,
Philippine ground orchid
Ground-dwelling orchids have swollen pseudobulbs found at or just below the soil level.

Range

Apart from the continent of Antarctica and other very dry and cold regions, orchids can be found growing naturally worldwide; from the damp mountainsides of Norway to the dry savannah of South America. The highest concentration of orchids is found in tropical regions.

Origins

The orchid family is entirely missing from the fossil record, which has caused much debate about its age. The discovery in 2007 of a bee preserved in amber carrying orchid pollen on its back indicates that orchids first appeared during the Late Cretaceous (80–76 million years ago).

Flowers

Orchids are known best for their unique flowers. While there is huge variation in color and size, they all conform to the same irregular pattern made up of three sepals and three petals. One of the petals is always different from the others, forming a labellum or lip, which gives an orchid flower its characteristic appearance. The remaining petals and sepals are usually very similar in size and shape. The labellum is typically much larger than the other petals and can be lobed, decorated with

Bulbophyllum anceps,
double-edged bulbophyllum
Modified aerial roots help epiphytic orchids cling
to their host and absorb moisture from the air.

Pollination

Bees, wasps, flies, ants, beetles, hummingbirds,
bats, and frogs have all been observed as the
pollinating agents of orchid flowers. Each species
of orchid usually has its own unique visitors.
The anatomy of the flower, as well as its scent
and color, will attract a particular insect to it.
A well-known mechanism is where a male bee is
fooled into thinking that the flower is a female
bee. The male bee will unsuccessfully mate with
the flower, and in doing so it transfers pollen
from one flower to the next.

plates, hairs, calluses, or keels, and frequently
bears bizarre color combinations. The flowers
of orchids make them easily distinguishable from
plants of other families, but there are also several
other characteristics that are unique to orchids.
For example, the seed from which a plant develops
is almost microscopic—made up of just a few
cells—and each fruit capsule from a single flower
can produce upward of one million seeds. These
tiny seeds require the help of soil-borne fungi if
they are to germinate and grow.

With its striking, slipper-shaped lip, the Queen's
Lady's-slipper orchid (*Cypripedium reginae*) nevertheless
conforms to the typical orchid flower structure.

Leaves

Many tropical and subtropical orchids possess
special water and nutrient storage organs called
pseudobulbs. These vary from slightly swollen
stems to shiny green organs from which the leaves
arise. The leaves themselves are usually fleshy,
simple in shape with smooth edges, and almost
always alternately arranged along the stem or
in a basal cluster.

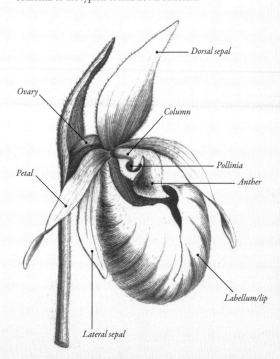

Dorsal sepal

Ovary

Column

Petal

Pollinia

Anther

Labellum/lip

Lateral sepal

Iridaceae
THE IRIS FAMILY

As seems appropriate for this family of very colorful plants, the name derives from the Latin word *iris*, meaning "rainbow." Practically all of these are herbaceous (nonwoody) and possess underground storage organs that are either corms (*Gladiolus*), rhizomes (*Sisyrinchium* and bearded iris), or, more rarely, bulbs (*Iris reticulata*).

Iris domestica,
blackberry lily

— *Rhizome*

— *Corm*

— *Saffron-covered filaments*

Range

With the exception of northern Asia, representatives of the iris family can be found worldwide, in both tropical and temperate regions. The main centers of diversity are South Africa, the eastern Mediterranean, and Central and South America.

Origin

Unfortunately the fossil evidence is very sparse when it comes to the iris family, so scientists are unclear as to its exact family history. There are distant ties with the *Orchidaceae*, and much closer links with the *Amaryllidaceae* and *Asphodelaceae*.

Size

There are 70 genera, containing more than 2,000 species. The family contains a good number of well-known garden plants, including *Crocosmia, Crocus, Dierama, Freesia, Gladiolus, Sisyrinchium, Tigridia*, and, of course, *Iris*. The plants have an upright habit and are reasonably low in stature, typically varying from ground level (*Crocus*) to no more than chest height (*Crocosmia*).

Crocus sativus,
saffron
Unknown in the wild, the saffron crocus probably derives from a spontaneous mutation of *C. cartwrightianus*, and has been propagated vegetatively for hundreds of years.

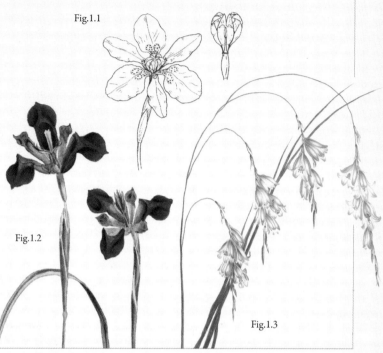

Fig.1.1. *Dietes butcheriana* (forest iris) is a rhizomatous plant that bears a close resemblance to the true irises (*Iris*). They differ from *Iris* in having tepals that are not joined to form a tube at their base.

Fig.1.2. *Moraea lurida* (cape tulip), like *Dietes*, is another iris lookalike with similar differences in the tepals. *Moraea* and *Dietes* are distinguished at the root: *Dietes* are rhizomatous; *Moraea* plants all grow from corms.

Fig.1.3. *Dierama pendulum* (wedding bell, or angel's fishing rod) is well described by its Latin name. *Dierama*, Greek for "funnel," describes the flower shape, and *pendulum* refers to the way they hang from the flower stem.

Fig.1.1

Fig.1.2

Fig.1.3

Flowers

While the general growth habit of the iris family is fairly consistent, there is a huge amount of variation in the flowers. Some are quite simple and regular, others more modified.

Consistent floral features include flowers with both male and female parts, six petals, and three stamens. Typically the flowers are arranged on long cymes or racemes, but sometimes there is just one flower on a single stem or, in the case of the crocuses, a single flower at ground level.

Arrangement of the six petals is either regular (*Sisyrinchium*) or irregular (some *Gladiolus*). They may unite at the base to form a tube of varying length, or more or less hang free from each other (*Moraea*).

They are borne in two whorls of three, which may be more or less equal in size—giving a flower of regular shape—or arranged differently, such as with bearded irises. With their three outer hairy petals and three inner upright petals, the petals of bearded irises are known as "falls" and "standards."

The *Iridaceae* are pollinated mostly by insects, such as butterflies, moths, beetles, and flies, but a few African species are adapted to pollination by sunbirds, including some of the genera *Crocosmia* and *Gladiolus*. Wind-pollinated *Dierama* (wand flowers) have pendulous, wiry-stemmed "fishing rod" inflorescences that sway in the breeze.

Gladiolus imbricatus,
Turkish marsh gladiolus

Amaryllidaceae

THE DAFFODIL FAMILY

Taking its name from the genus *Amaryllis*, the *Amaryllidaceae* is perhaps better known for daffodils (*Narcissus*). This is a family of herbaceous perennials that produce strap-shaped leaves and heads of flowers from an underground bulb or rhizome. The plants typically emerge in spring and die back after flowering.

Size

This is a fairly large family of 600 species across 90 genera. Since the description and classification of *Amaryllidaceae* at the start of the 19th century, there has been a lot of reorganization, with some genera moved and reclassified. *Hippeastrum*, for example, has been broken into a number of different genera, such as *Habranthus*, *Pyrolirion*, *Zephyranthes*, and *Sprekelia*. Three subfamilies are recognized: the *Agapanthoideae*, containing *Agapanthus*; the *Allioideae*, with the onions (*Allium*); and the *Amaryllidoideae*, home to the daffodils (*Narcissus*) and snowdrops (*Galanthus*).

Nerine humilis,
dwarf nerine

Narcissus major,
great daffodil

Range

This family is found worldwide, but mainly in warm temperate, subtropical, and tropical regions. *Narcissus*, *Galanthus*, and *Leucojum* (summer snowflake) have a more northerly distribution, stretching into cool temperate northern Europe.

Origin

The fossil record of *Amaryllidaceae* is very poor, so very little is known about its evolutionary history. While botanists suspect the origins to be in the Cretaceous, the earliest fossil records are from the much later Miocene epoch, just 15 million years ago.

Flowers

The showy, three-parted flowers are bisexual and almost always regular in shape. They are carried in umbellate flower heads, sometimes singly or in twos or threes. The emerging flowers or flower heads are usually wrapped in a papery or membraneous bract.

Since the petals and sepals are all very alike, they are referred to jointly as tepals. These form two whorls of three (six tepals altogether), which are either free from each other (*Galanthus*, *Leucojum*, *Amaryllis*, *Nerine*) or joined to form a tube or funnel (*Crinum, Zephyranthes, Sternbergia, Cyrtanthus, Stenomesson*).

The flowers of *Narcissus* have a very distinctive, extra inner structure known as a corona. This can be enlarged like a trumpet or reduced to a cup or disc, and it is often a contrasting color to the corolla. Corona-like structures can also be seen in some *Allioideae* genera (*Pancratium, Hymenocallis, Tulbaghia*).

Ovaries are superior in *Agapanthoideae* and *Allioideae* subfamilies, and inferior in *Amaryllidoideae*. They are made up of three fused carpels and ripen into a dry capsule or fleshy fruit, as seen in the genus *Clivia*.

Leaves

Most members possess linear or strap-shaped leaves emerging from the bulbous or rhizomatous root stock. They are typically deciduous with a few evergreens (*Clivia*). The foliage is fleshy and tender rather than fibrous.

USES FOR THIS FAMILY

This family contains many ornamental plants; as well as daffodils, snowdrops, and snowflakes, there are belladonna lilies (*Amaryllis belladonna*), *Hippeastrum*, and late-flowering *Nerine* and *Sternbergia*. Genus *Allium* contains some very important vegetables, such as onions, leeks, garlic, and chives.

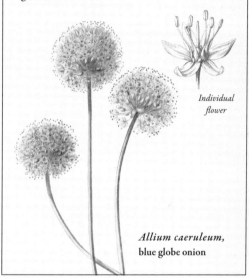

Individual flower

***Allium caeruleum*,**
blue globe onion

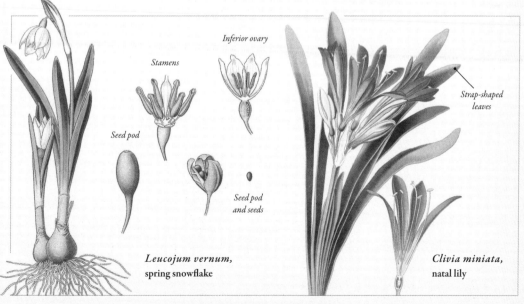

Inferior ovary

Stamens

Seed pod

Seed pod and seeds

***Leucojum vernum*,**
spring snowflake

Strap-shaped leaves

***Clivia miniata*,**
natal lily

Asphodelaceae
THE DAYLILY FAMILY

This collection of perennials and the odd tree or shrub has recently been assembled and includes plants from several other families. Previously known as *Xanthorrhoeaceae,* the name was changed, making it easier to pronounce and spell.

Size

There are around 900 species in the family with more than half in the succulent genus *Aloe*. Other large genera include *Bulbine* (80 species), *Kniphofia* (75 species), and *Haworthia* (60 species). *Hemerocallis* (daylily), *Phormium* (New Zealand flax), and *Eremurus* (foxtail lily) have ornamental value in the garden.

Range

Most members of this family are restricted to the Old World (Asia, Africa, and Europe), both temperate and tropical, with only two genera present in South America. Southern Africa and Australia are especially rich in species, while the family is absent from the polar regions and North America.

Origins

Very little fossil evidence exists for the family, with an isolated specimen from the Eocene (around 45 million years old) deposits found in Australia.

Flowers

In the most familiar species, flowers are borne on a stalk, branched or unbranched, held above the leaf rosette. The blooms are often colorful, with six similar sepals and petals. These can be fused into a tube, as in *Kniphofia* (red-hot poker), partially fused (daylily), or entirely free (asphodel, *Asphodelus*). With *Aloe*, the sepals are partially fused, while the petals are free. Several daylily cultivars have multiple petals. There are six stamens per flower, occasionally hairy, with copious nectar in some blooms.

Hemerocallis dumortieri, daylily

Bulbine alooides, aloe-like bulbine

Pollination

Many *Asphodelaceae* are bird-pollinated. These are easily identified by their red or orange tubular flowers with leathery petals, profuse nectar, and lack of scent. They also have strong flower stems because "Old World" bird pollinators cannot hover like "New World" hummingbirds, and need a perch.

USES FOR THIS FAMILY

Herbaceous borders would not be the same without daylilies; they are floriferous, reliable plants available in many colors. With more than 70,000 known daylily cultivars, and more on the way, there is one for every location. Other useful herbaceous perennials include red-hot pokers and asphodels, while the succulent genera make excellent houseplants. Try growing *Aloe aristata*, *A. striatula*, and *A. polyphylla* in patio pots.

Kniphofia triangularis,
dwarf red-hot poker

Dianella caerulea,
flax lily

Fruit

Fruits are typically dry capsules, though the bright blue-to-purple berries of *Dianella* are especially attractive.

Leaves

Usually arranged in rosettes, the leaves are alternate or sometimes two-ranked (*Aloe plicatilis*), have no distinct petiole, and can be armed with marginal spines. In several genera (*Aloe, Haworthia, Gasteria*), the leaves are succulent, while in others (*Phormium, Dianella*), the leathery leaves appear folded together at the base, opening out further up the plant. While most species are herbs, there are a few trees (*Aloe dichotoma, Xanthorrhoea*) and climbers (*A. ciliaris*).

Asparagaceae
THE ASPARAGUS FAMILY

Previously a minor family, or even a mere division of the *Liliaceae* (see pages 78–79), changes in taxonomy have greatly expanded the *Asparagaceae*, making it one of the biggest in the monocots. Well-known families included here are *Agavaceae*, *Convallariaceae*, *Hyacinthaceae*, and *Ruscaceae*.

Size

With upward of 2,250 species, this large family encompasses numerous forms including trees (*Dracaena*, *Yucca*), succulents (*Agave*, *Sansevieria*), vines (*Asparagus*, *Semele*), perennials (*Polygonatum*, *Hosta*), and bulbs (*Hyacinthus*, *Camassia*).

Range

Distributed almost worldwide, though not in the polar regions, the succulent species are typically found in deserts and other arid habitats.

Convallaria majalis,
lily-of-the-valley

Origins

As with many other monocot families, the fossil record is sparse. Fossilized leaves similar to those of modern cabbage palm (*Cordyline*) have been found in Australia, dating from the Eocene (56–34 million years ago).

Flowers

Featuring many different flower forms, this is a challenging family to define. All typically have six similar sepals and petals that can be free (asparagus, spider plant), partially fused (hyacinths, bluebells), or fully fused (*Convallaria*, lily-of-the-valley). Six stamens are usually present and they can be free, fused to the petals, or fused into a tube. Bracts, such as those on asparagus spears, are often present on the inflorescences (*Camassia*, *Beschorneria*).

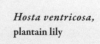

Hosta ventricosa,
plantain lily

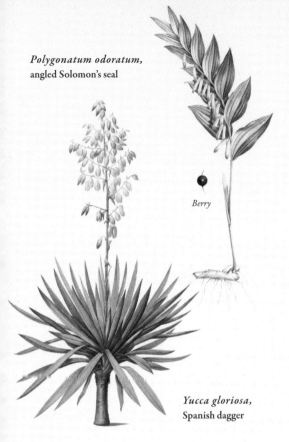

Polygonatum odoratum,
angled Solomon's seal

Berry

Yucca gloriosa,
Spanish dagger

Fruit

Fruits are commonly capsules (*Hosta, Yucca*) or berries (*Ruscus, Polygonatum*), containing black (or, rarely, brown) seeds.

Leaves

It is to be expected that such a diverse family would have many varieties of leaf form. Leaves are either clustered in rosettes or alternate along stems, rarely two-ranked. Typically they are entire and without petioles (with the exception of *Hosta*) and can be leathery, succulent, or papery in texture. Many century plants (*Agave*) have spiny leaf margins. In butcher's broom (*Ruscus*) and asparagus, the leaves are reduced to scales, and flat or needlelike stems, known as phyllodes, take on the role of photosynthesis.

USES FOR THIS FAMILY

Other than ornamentals, few *Asparagaceae* are put to great use by people. *Agave* species are the source of fibers such as sisal (*A. sisalana*) and the liquor tequila (*A. tequilana*), while some bulbs are edible when prepared correctly (*Camassia, Ornithogalum*). But perhaps the best-known product is asparagus itself, and in spring, the newly harvested spears fetch a premium. They resemble the shoots of other family members, including *Agave, Ruscus,* and *Camassia*; a useful feature in an otherwise tough-to-identify family.

Some of the toughest (and most frequently abused) houseplants can be found in *Asparagaceae* including cast-iron plant (*Aspidistra*), mother-in-law's tongue (*Sansevieria*), dragon tree (*Dracaena*), and spider plant (*Chlorophytum*). For spring color in the yard, plant bulbous camas lilies (*Camassia*), hyacinths (*Hyacinthus, Muscari*), and squills (*Scilla*), while useful ground covers include lily-turf (*Liriope*), mondo grass (*Ophiopogon*), and Nippon lily (*Rohdea*).

Asparagus officinalis,
common asparagus

Arecaceae
THE PALM FAMILY

The quintessential tropical plant, palms are a conspicuous component of their environment. They are usually single-stemmed trees topped with a crown of leaves, but some cluster or branch to form shrubs or vines.

Size

There are around 2,400 palm species. Well-known examples are coconut, date, and oil palms, and to a lesser extent, sago, betel-nut, and the rattans from which cane furniture is constructed.

Range

While palms stretch from Mediterranean France in the north, to New Zealand in the south, they are primarily a tropical family. They can be found on every continent except Antarctica.

Cocos nucifera,
coconut palm

Flowers

The basic bloom of palms is not highly ornate or decorative. They typically have three small sepals, three small petals, and six stamens, and usually lack vivid color or scent. Some palms make up for this deficiency by producing lots of flowers; the inflorescence of talipot palm (*Corypha*) is the largest of all plants, with more than 23 million individual flowers.

Origins

With their tough leaves and trunks, palms have preserved well as fossils, providing a record dating back at least 80 million years to the Upper Cretaceous.

Corypha taliera,
talipot palm
The resources required to produce the world's largest flowering structure are so great that this palm dies after fruiting.

Fruit

Palm fruits range from bright red, orange, and yellow to glossy black. Rattan fruits are covered in unique, almost reptilian scales, while others bear spines, hairs, or warts. The seaborne fruits of coconut have fibrous husks for flotation. Within each fruit are one or more seeds, sometimes protected by a hard outer layer. These range in size, the smallest being less than half an inch and the largest being double coconut (*Lodoicea*), which at 12 inches long and 55 pounds in weight is the world's largest seed.

Leaves

The resplendent leaves in the palm family bring a touch of tropical luxuriance to a garden, and are composed of three sections: sheath, petiole, and blade. The sheath wraps around the stem and, when the leaf falls, a distinctive scar remains on the trunk. In other palms, fibrous sheaths are retained, giving a shaggy impression. Petioles have spiny, sharp, or unarmed margins. The blade is typically divided into leaflets with feather-like (known as pinnate) or handlike (known as palmate) arrangements. Palms with pinnate leaves have short petioles and numerous individual leaflets; palms with palmate leaves have long petioles and leaflets that join at the base to form a fan.

USES FOR THIS FAMILY

Florida, California, and the Gulf Coast are home to many palm trees (native and non-native). In more temperate areas, the chusan palm (*Trachycarpus fortunei*), native to the Himalayas, will emerge largely unscathed from most winters. The more compact European fan palm (*Chamaerops humilis*) is another hardy species, but only where sharp drainage is assured, and the jelly palm (*Butia capitata*) will also thrive in colder climes, especially coastal areas. Sabal palm is native to the United States and is common in the Southeast. Many palms make perfect houseplants, including parlor, kentia, and white elephant palm (*Chamaedorea elegans, Howea forsteriana, Kerriodoxa elegans*).

Trachycarpus fortunei,
chusan palm

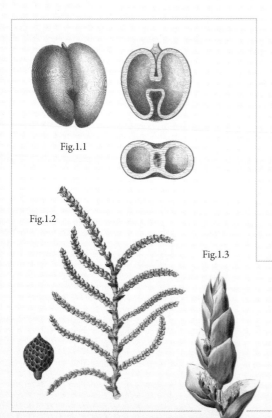

Fig.1.1

Fig.1.2

Fig.1.3

Fig.1.1. *Lodoicea maldivica* produces the world's largest seed, pictured here whole and in section.

Fig.1.2. *Calamus ornatus* is a climbing rattan palm with small flowers and scaly fruits.

Fig.1.3. *Plectocomia elongata,* another climbing rattan, produces flowers within enlarged bracts.

Zingiberaceae
THE GINGER FAMILY

Consisting entirely of herbaceous perennials, plants in the ginger family are rich in aroma. Cardamom, turmeric, galangal, and grains of paradise, along with the eponymous ginger, are all harvested from *Zingiberaceae*.

Size

In tropical areas the 1,275 or so species make for important garden plants; for example: torch ginger (*Etlingera*), golden brush (*Burbidgea*), shell ginger (*Alpinia*), and dancing girl (*Globba*). In cooler regions, garland flower (*Hedychium*), *Roscoea*, and *Cautleya* can be grown, though winter protection is often necessary.

Range

Zingiberaceae can be found in most tropical regions either in marshland or, more often, in the forest understory. Outside the tropics, species from more elevated areas are preferred for cultivation.

Origins

The oldest ginger fossils date to around 85 million years ago in the Late Cretaceous. Younger fossils found in Europe and North America date from a time when these continents enjoyed a warmer climate. *Zingiberaceae* no longer occur naturally in either region.

Flowers

The exotic blooms of *Zingiberaceae* appear at the tips of leafy shoots, or direct from the rhizomes, with individual flowers emerging from within clusters of bracts. In some species, the bracts are highly colorful and often more attractive than the actual flowers. Each flower has three sepals fused into a tube, and three petals that are partially fused, with one lobe often larger than the other two. The often unremarkable petals are outshone by two to four petal-like staminodes (modified stamens), with the inner two highly colored and sometimes fused together into a lip. There is only one fertile stamen in each flower.

Alpinia nutans,
dwarf cardamom
Highly modified stamens, known as staminodes, form the decorative lower lips in the flowers of many ginger species.

*Capsule in
section*

*Capsule split open,
with seed*

Globba radicalis,
dancing girl ginger

Roscoea,
alpine ginger

Fruit

The fruits are capsules, which may be dry or
fleshy at maturity. The seeds may have a
colorful fleshy coating.

Leaves

Ginger leaves are simple, entire, alternate, and
usually arranged in two ranks. They often have
distinct petioles, with grasslike ligules at the
base. The petioles open out to form sheaths and
these overlap, forming the upright
"stems." The actual stems are the
rhizomes that grow on the surface
of the soil, or just below it.

Zingiber officinale,
ginger

93

Bromeliaceae
THE PINEAPPLE FAMILY

This group of herbaceous perennials rarely appears in yards in the northern hemisphere, though not for their lack of color or architectural poise. This family is very common in Florida, along the Gulf Coast, Texas, and California. Spanish moss, the trademark of Southern gardens, from Charleston to New Orleans, is a bromeliad.

Size

Of the nearly 2,650 species, only pineapple (*Ananas comosus*) is a cultivated crop, but decorative plants abound, including urn plant (*Aechmea fasciata*), queen's tears (*Billbergia nutans*), and the many air plants (*Tillandsia*).

Range

Almost entirely restricted to the Americas, bromeliads range from Florida to Argentina, with one species, surprisingly, in West Africa (*Pitcairnia feliciana*). Most are denizens of tropical rain forest, though a handful of genera prefer the desert and a few thrive high in the Andes.

Origins

Few fossilized bromeliads have been found and it seems likely the family is of recent origin.

Flowers

One of the most appealing decorative features of *Bromeliaceae* is that their flowers are often accompanied by vividly colored bracts. These modified leaves last long after the flowers fade, greatly prolonging the display. Inflorescences appear from the center of the leaf rosette; some are tall and imposing, while others barely emerge at all. The actual flowers are often small, and each has three sepals, three petals, and six stamens. In many bromeliads, flowering terminates the life of the individual leaf rosette.

Tillandsia cyanea,
pink quill
This air plant's blue flowers may be short-lived, but the pink bracts surrounding them can last for several weeks.

Ananas comosus,
pineapple
The many flowers in this
inflorescence each form their
own fleshy fruits, and these in
turn merge together to form
a single pineapple.

Fruit

Fruits are often dry capsules,
releasing seeds with silky hairs to aid
wind dispersal. By contrast, in the
pineapple, each flower in the
inflorescence becomes a fleshy fruit,
eventually merging together to produce
a large compound fruit. Each scale on
the pineapple skin represents a fruit
derived from one flower.

Leaves

Bromeliad leaves are alternate, often strap-shaped,
and lack petioles. They generally form a rosette,
which in some species collects rainwater, leading
to names such as urn plant. Some bromeliads,
especially those from arid regions, have
spines along the leaf margins and many
have prominent veins or brightly
colored patches in the center of the
rosette. Once a rosette has flowered,
it will often die, though new rosettes
form around the base of the plant.

Tillandsia aeranthos,
air flower
All air plants grow as epiphytes, attaching
themselves to tree branches, rocks, and even
telegraph poles and cables.

Where to water

Most bromeliads from tropical
forests are epiphytes, which
means they grow by attaching
to tree branches, never in contact
with the soil. They are not parasites
and do not take from the tree; it
merely provides a perch in the sunny
upper canopy. The urn-shaped leaves of
many species help them to avoid drought
and, when growing them indoors, it's
important to regularly top up their water
supply. Air plants adopt a different
strategy; instead of forming urns, their
leaves are covered in silver scales that
rapidly absorb rainwater, so these
plants should be doused.

The pineapple and many
desert bromeliads do not
form urns and should be
watered at the roots.

Poaceae
THE GRASS FAMILY

The grass family is one of the most prominent in the Earth's ecosystems and to humans it is economically by far the world's most important group of plants. It includes the bamboos, as well as all the grain crops such as rice, corn, and wheat.

Size

There are more than 10,550 species of *Poaceae*, split across approximately 115 genera of bamboos and 600 genera of true grasses. Strictly speaking, they are all nonwoody annuals or perennials, but the question of woodiness is relative when it comes to the canes of bamboo. While they are not truly woody in the same way as dicot trees and shrubs, some canes rival the strength of steel.

Oryza sativa,
rice

These two large groups are further broken down in various ways. Bamboos tend to be split into nonwoody bamboos, tropical woody bamboos, and temperate woody bamboos. Grasses can be split either by way of growth habit (clump-forming, spreading, or carpeting) or into "cool season" and "warm season" types.

Range

The family exists in almost every ecological niche from pole to pole, and from mountain top to the shore of every continent. It is truly cosmopolitan, forming an estimated 20 percent of the Earth's vegetative cover.

The great grasslands—the Asian Steppes, South American Cerrado, North American Prairie, and African Savannah—are vast and important habitats, rivaling the tropical rain forests in terms of their biodiversity and land coverage. These habitats occupy a climatic zone between forest and desert. In Antarctica there are only two species of flowering plant; one of these is a grass.

Origin

Analysis of fossilized dinosaur dung shows that grasses were part of the dinosaur diet about 100 million years ago. Animals with teeth that were specialized for grazing grasses appear in the fossil record about 80 million years later,

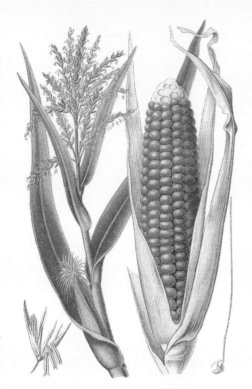

Zea mays,
corn
A strongly upright plant with leaves growing from each node, a few that form inflorescences tightly wrapped in many leafy husks. On fertilization, these swell into corn cobs.

suggesting that in this time grasses had begun to dominate many habitats.

The earliest grasses and bamboos grow on the tropical forest floor where there is little wind. These primitive grasses are pollinated by insects. The vast majority of grasses, however, are wind-pollinated, and this was probably an adaptation that served grasses on the forest fringes well, helping them to diversify and spread.

The relationship of grasses to other plant families is not entirely clear. It might appear that the sedges and rushes are near relatives, but their close resemblance is most likely to be a result of parallel evolution, rather than a recent shared ancestry.

USES FOR THIS FAMILY

In the form of the grain crops rice, wheat, maize, oats, and barley, the *Poaceae* is central to feeding the world, and a major part of the economy. Bamboos make valuable construction materials, while much of the world's sugar comes from the tall grass *Saccharum officinarum* (sugarcane).

Grasses, in the form of lawns and sports turf, have a significant role for both landscaper and gardener. Grasses and bamboos also have excellent architectural properties. Unfortunately, there are also many weedy species that are spread by seeds or creeping roots, some of which are very persistent.

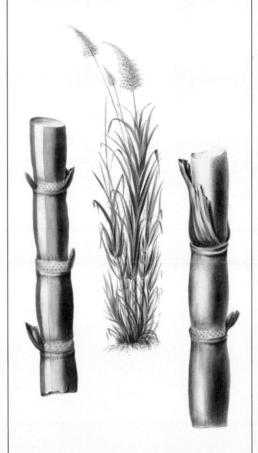

Saccharum officinarum,
sugarcane

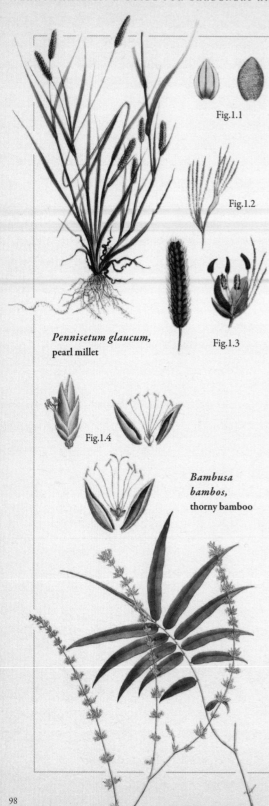

Fig.1.1

Fig.1.2

Pennisetum glaucum,
pearl millet

Fig.1.3

Fig.1.4

*Bambusa
bambos,*
thorny bamboo

Flowers

Grasses are identified by their tiny, highly modified flowers arranged in racemes, spikes, or panicles. The individual flowers are so specialized, it is as if the whole floral mechanism has been stripped down and reinvented using just the bare essentials.

The petals are reduced to small structures called lodicules (sometimes even these are absent), and there are usually just three stamens and two stigmas. These are all sandwiched between protective scalelike structures called the palea and lemma, which are highly modified versions of the sepals.

The function of the lodicules is to swell with water and force apart the flower so that the stamens and stigmas are exposed for pollination. The anthers are attached by their middles to the filaments so that they rock in the wind and release their pollen. The stigmas are feathery, so the chance of catching the windborne pollen is maximized.

Fig.1.1. The palea and lemma on pearl millet (seen here, L–R) are scalelike structures that resemble and derive from sepals. Their job is to protect the flower parts within.

Fig.1.2. Feathery, threadlike stigmas protrude from the open flowers of peak millet, to stand the greatest chance of capturing the windborne pollen.

Fig.1.3. The narrow, cattail inflorescences of *Pennisetum* are densely packed with many flowers, interspersed with bristles.

Fig.1.4. This flower detail of *Bambusa bambos* shows closed flower (left) and individual flower parts (below and right), with protective lemma and palea, and stamens and stigmas within.

Close inspection of grass flower heads reveals the beautiful subtleties in their variation and texture, which can be put to great use in planting schemes. From the foxtail spikes of the fountain grasses (*Pennisetum*) to the feathery plumes of pampas grasses (*Cortaderia*), and from the airy panicles of moor grass (*Molinia*) to the whiskers of barley (*Hordeum*), there is huge variation.

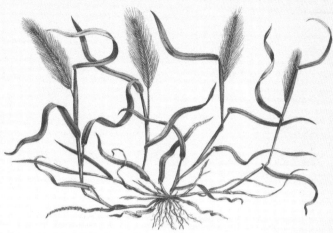

Hordeum murinum,
barley

Leaves

Grass leaf blades are typically long and narrow, but they can be wider in tropical species and in those that grow in shade. There are two parts to a grass leaf: the blade and the sheath, the latter of which envelops the stem and provides mechanical support. The sheath also protects the unspecialized cells that lie around the node.

The small outgrowth at the base of many grass leaves is called a ligule and is an important feature in grass identification.

Along the stems are small swellings called nodes, to which the leaves are attached. The nodes contain unspecialized cells that allow the stem to bend upright again after it is trampled or flattened. These cells also allow leaves to continue growing after they are cut or grazed.

The fact that grasses can regenerate from the node and leaf base means that they can tolerate the extreme grazing and trampling that would eliminate most other plants. This innovation in anatomy is the main reason for the huge success of grasses.

Roots

The root system is fibrous, readily forming adventitious roots from the lower part of the stem to form a tussock. Some species are able to extend sideways by underground rhizomes or surface stolons to form a close sward (as in a lawn) or an impenetrable thicket (as with some bamboos).

Paspalum
fimbriatum,
Columbia grass

Malus coronaria,
sweet-scented crab apple
From the rose family
(*Rosaceae*—see pages 130–133)

CHAPTER 3

Eudicots

About 85 percent of flowering plants are eudicots. They make a substantial contribution to our yards with, among others, roses, sunflowers, clematis, maples, rhododendrons, and geraniums. They're easily distinguished from monocots by their branching leaf veins and flowers with parts in fours or fives (though there are a few exceptions).

This sizeable group can be loosely subdivided into three major lineages, the smallest of which is the early eudicots, which includes poppies, buttercups, and others, encompassing a huge diversity of forms, from trees and shrubs, to vines and aquatic plants. The remaining eudicots split almost in half; those related to roses, known as superrosids, and those related to daisies, known as superasterids. These two super-groups can be tricky to identify, though superrosids often bear stipules, while superasterids typically have tubular flowers; both provide an abundance of plants suitable for garden use.

Well-known superrosids include edible apples, oranges, strawberries, cabbages, and beans; ornamental trees such as oaks, maples, birches, and rowans; plus important flowers such as peonies, violets, fuchsias, and lupines. Superasterids include flowering shrubs like hydrangeas, camellias, hebes, and heathers; blooming vines such as morning glory, honeysuckle, and jasmine; not to mention edible parsley, rosemary, potatoes, and tomatoes.

Berberidaceae

The barberry family

This is a comparatively small family, but one of great diversity and importance in the garden. Barberry (*Berberis*), Oregon grape (*Mahonia*), and heavenly bamboo (*Nandina*) are shrubs, though most species are herbaceous, including barrenwort (*Epimedium*) and May apple (*Podophyllum*).

Size

While there are many ornamentals, the 715 or so species of *Berberidaceae* do not include any crops. A few *Berberis* are quite invasive in the United States so their sale and cultivation is controlled.

Podophyllum peltatum,
May apple

Range

Most genera are restricted to eastern Asia and North America, but the barberries also extend across Europe into North Africa, and along the Andes into South America.

Origins

Considered one of the earliest of the dicot families, fossilized pollen from the Cretaceous (around 90 million years ago) has been tentatively identified as *Berberidaceae*.

Flowers

Berberidaceae flowers show several primitive features; sepals and petals can be absent (*Achlys*) or numerous, and often undifferentiated, though they are usually in whorls of two or three. In many barrenworts, the innermost petals form elaborate horn-shaped structures. Each flower has six stamens or, in rare cases, either four (*Epimedium*) or more (*Podophyllum*).

Berberis aggregata,
salmon barberry

USES FOR THIS FAMILY

Barberries are widely seen in parking lots and other municipal settings. Many are extremely invasive, including *Nandina* species. *Mahonia* are much better ornamental choices for the garden. Their flowers and berries are vibrant, as is the fall foliage of some deciduous species, and throughout the summer many cultivars of Japanese *Berberis thunbergii* have richly colored foliage. Barberries make almost impenetrable hedges and are one of the few woody plants to be shunned by hungry deer. The often more glamorous *Mahonia* provides a year-round foliage display and useful winter blooms, while herbaceous and evergreen barrenworts will thrive in difficult dry shade. The dainty divided leaves of *Vancouveria* and the bold spotted leaves of *Dysosma* are ideal in a woodland garden.

Nandina domestica, heavenly bamboo

Fruit

Typically the fruits of *Berberidaceae* are fleshy berries (*Berberis, Mahonia, Podophyllum, Nandina*) or dry follicles (*Epimedium, Vancouveria*).

Leaves

In such a variable family, little can be said about the foliage as a whole; it is best therefore to adopt a genus-by-genus strategy. Barberries are either evergreen or deciduous, with simple leaves and spiny stems. Evergreen *Mahonia* has compound leaves with a feather-like (pinnate) arrangement of often spiny leaflets. While these two seem rather different, many botanists lump them together into one genus: *Berberis*. Both generally have vivid yellow wood beneath the bark. Sacred bamboo has finely divided, unarmed foliage that can turn bright red in the fall. In herbaceous *Podophyllum* and *Diphylleia*, the leaves are maplelike, while in

others, they are compound with two (*Jeffersonia*), three (*Achlys, Ranzania,* some *Epimedium*) or more leaflets. Most species have alternate leaves (those of May apple, however, are opposite) with toothed or lobed edges.

Mahonia napaulensis, Nepal mahonia

Papaveraceae
THE POPPY FAMILY

The childlike simplicity of the poppy flower is so much a part of visual culture in the West that it cannot be mistaken for any other flower. However, family representatives come in several colors and guises, such as *Fumaria*, *Lamprocapnos*, and *Macleaya*, which have flowers that look very different from true poppies.

Size

With 43 genera and roughly 820 species, the poppy family is mostly made up of herbaceous (nonwoody) annuals, biennials, and perennials. There are a few woody or shrubby representatives such as tree poppy (*Dendromecon*) and Californian tree poppy (*Romneya*). There are about 80 species of true poppy (*Papaver*).

Most authorities place genera such as *Fumaria*, *Corydalis*, and *Lamprocapnos* within *Papaveraceae*, but some separate them into *Fumariaceae*. They have irregularly shaped flowers and compound, almost feathery foliage.

Range

People from tropical regions may have never seen a poppy because it is a plant almost exclusive to temperate regions in the northern hemisphere. One outlying exception is the South African poppy, *Papaver aculeatum*. Some of the Arctic poppy species are among the most northerly growing land plants.

Origin

Fossils of the earliest known species, the Late Cretaceous poppy *Palaeoaster inquirenda* (now extinct), date from 75 million years ago. Specimens have been found at numerous excavation sites; in South Dakota, one was found close to the remains of a *Tyrannosaurus*.

Flowers

Large, bright, and conspicuous, the flowers of poppies are hard to ignore, whether they are growing in open meadows, garden borders, or on roadside edges. They carry both male and female parts and—with some notable exceptions, such as *Lamprocapnos*, *Fumaria*, and *Corydalis*—have a regular, symmetrical shape.

Papaver dubium,
long-headed poppy

USES FOR THIS FAMILY

Besides the true poppies, there are several other genera of horticultural merit with very similar flowers, such as Himalayan blue poppies (*Meconopsis*), horned poppies (*Glaucium*), prickly poppies (*Argemone*), and Californian poppies (*Eschscholzia*).

Other ornamental species include *Corydalis* and *Fumaria*, both of which are pretty border perennials with feathery foliage. The plume poppy (*Macleaya*) is a tall, architectural plant that bears voluminous, airy flower heads made up of many tiny flowers. The bleeding heart (*Lamprocapnos spectabilis*) is a popular border perennial that, until recently, was known as *Dicentra spectabilis*.

A blue Himalayan poppy of the genus *Meconopsis*

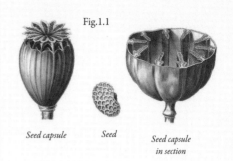

Fig.1.1

Seed capsule Seed Seed capsule in section

Fig.1.1. Like other poppies in the genus *Papaver*, the opium poppy (*Papaver somniferum*) bears inflated "pepper pot" seed capsules that only release the seed when dry.

Fig.1.2. The very bright and colorful flowers of Iceland poppies (*Papaver nudicaule*) come in many colors. The plants make very popular short-lived bedding plants.

Fig.1.3. The distinctive flowers of bleeding heart (*Lamprocapnos spectabilis*) are formed by the inner white petals protruding from the outer pink petals. They are carried on arching racemes.

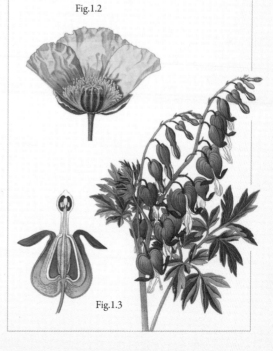

Fig.1.2

Fig.1.3

An unusual feature of the flowers are the two green sepals; these enclose the flower but drop off unnoticed just before the flower opens. There are usually four petals in two whorls, and they are often crumpled in the bud. Petals are absent in *Macleaya*.

The "pepper pot" seed head that soon follows the flower is a characteristic of some genera of the family. When dry, these open by valves or pores to release their many tiny seeds when shaken by the wind or by passing animals.

Ranunculaceae
THE BUTTERCUP FAMILY

With the exception of the woody climbers *Clematis* and *Clematopsis*, this is a family of nonwoody, herbaceous plants, most of which are perennial, with an odd sprinkling of annual plants such as larkspur (*Delphinium*) and love-in-a-mist (*Nigella*). There are no buttercup trees.

Size

With more than 1,800 species, the buttercup family contains a good number of very well-known wild flowers and garden ornamentals, such as anemones, clematis, hellebores, and, of course, the buttercup. It also contains some very poisonous plants, such as the monkshood (*Aconitum*).

Ranunculus acris,
common/meadow
buttercup

Range

This is an enormously successful family of plants, with family members spread across almost all four corners of the world. Perhaps one of the reasons that this family of plants is so well known to gardeners is because the majority of the *Ranunculaceae* can be found in temperate and cold regions of the northern hemisphere.

Origins

The *Ranunculaceae* is generally regarded as a primitive family, one of the first flowering plant families to evolve during the Early Cretaceous, long before more advanced plants such as grasses (*Poaceae*).

Flowers

The buttercup itself gives us lots of clues about the characteristics of the *Ranunculaceae* family. The flowers are fairly simple and radially symmetrical, but it is one of the few species where the petals and sepals (collectively known as the calyx) are almost indistinguishable; some botanists refer to them as tepals.

Another example is clematis: the flower bud is enclosed by the sepals, which protect the inner workings of the flower. As it grows and expands,

Clematis jackmanii,
Jackman's clematis

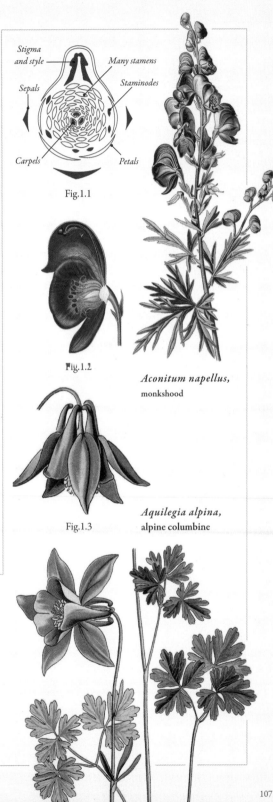

Fig.1.1

Aconitum napellus,
monkshood

Fig.1.2

Fig.1.3

Aquilegia alpina,
alpine columbine

the sepals open up and become much more colorful, just like the petals within.

Because the buttercup flower is so recognizable, it almost blinds us to the true diversity of this fascinating family of plants. Not all the flowers are round, radially symmetrical, and regular, and there are many other oddities besides. Take, for example, the hooded flowers of *Delphinium* and monkshood (*Aconitum*). Evolution has caused the upper sepals of these species to become enlarged and curved over—most likely as a strategy to attract particular pollinating insects. Spurred flowers are also sometimes seen, most notably in the distinctive blooms of the columbine (*Aquilegia*).

Fig.1.1. This diagram shows an aerial view of monkshood flower parts arranged in radial symmetry, typical of the buttercup family, but with an elongated upper petal and anthers.

Fig.1.2. In cross-section, the monkshood's affinity to the buttercup is more easily seen.

Fig.1.3. The spurred flowers of the columbine may be unique, but their structure is simply a variation on the buttercup theme.

Pollination

The flowers of the buttercup family show many of the hallmarks of a primitive ancestry: there is a multitude of flower parts, which are spirally arranged and with a superior ovary. This makes pollinating easy for beetles, which are probably the earliest pollinators of flowers. Complex flower forms and scents are largely absent, with the exception of the previously mentioned delphiniums, monkshoods, and columbines.

Pollinating insects are drawn into the center of the flower, where they will find a reward of pollen or nectar within. The genera *Anemone*, *Pulsatilla*, and *Clematis* do not produce nectar and are visited only for their pollen. Usually it is the colorful petals or sepals that attract the insects in the first place but, in the case of some *Thalictrum* species, the attractive stamen filaments or anthers do the job in unison with the smaller tepals.

Leaves

In most *Ranunculaceae*, the leaves are borne in one of two ways: they are either basal (meaning they grow from the base) or they are caulescent, which means that they are carried up the growing shoot. Those that grow up the stem are usually alternately arranged, although *Clematis* are a notable exception, with the leaves arranged in opposite pairs—a feature readily apparent to anyone familiar with pruning these climbing plants.

Ranunculaceae leaves are commonly much-divided (palmately lobed), but of course there are many exceptions, such as the heart-shaped leaves of the lesser celandine (*Ranunculus ficaria*) or the enormously feathery leaves of the water crowfoot (*R. aquatilis*), developed as an adaptation to their free-flowing, watery habitat.

Some flowers are surrounded by an involucre (a whorl of bracts beneath a flower), as seen in some anemones, pasque flowers (*Pulsatilla*), and, most notably, the curious flowers of love-in-a-mist (*Nigella*). These are sometimes mistaken for being part of the flower structure, but involucres are actually modified leaves.

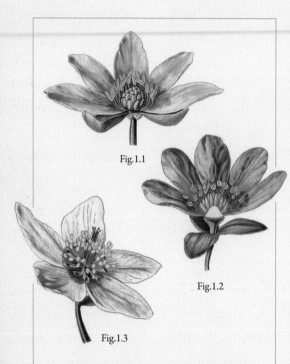

Fig.1.1

Fig.1.2

Fig.1.3

Fig.1.1. Collectively known as the calyx, the sepals and petals of lesser celandine (*Ficaria verna*) are similar in form and sometimes indistinguishable.

Fig.1.2. The leafy involucre of the pasque flower (*Pulsatilla vulgaris*) is often mistaken for the sepals, but the true sepals in fact belong to the calyx.

Fig.1.3. The color and pattern of the calyx is vital to *Helleborus niger* (Christmas rose) for pollination.

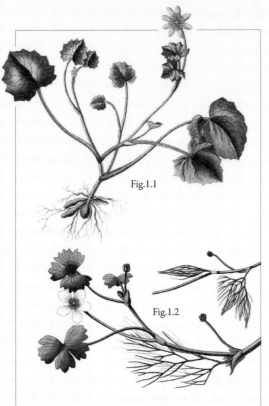

Fig.1.1

Fig.1.2

Fig.1.1. The heart-shaped leaves of the lesser celandine (*Ficaria verna*) emerge in late winter and the early flowers are a harbinger of spring.

Fig.1.2. The feathery leaves of the water crowfoot (*Ranunculus aquatilis*) are so different from other *Ranunculus* species that some botanists believe they should belong to a separate genus, *Batrachium*.

Fig.1.3. *Nigella* is an exceptional member of the buttercup family, and not just for its fancy involucres: the plants are also annuals.

Fig.1.3

USES FOR THIS FAMILY

The *Ranunculaceae* has a plant for almost every condition and aspect. Delphiniums for full sun, water crowfoot for streams, kingcups (*Caltha palustris*) for ponds, hellebores for winter, aconites for spring, and anemones for shade. The clematis comes into its own in yards because it is a climber and will colonize vertical spaces. Then there is the popular, late summer-flowering Japanese anemone. A number of genera are highly poisonous and have caused deaths. Victorian medical books give lurid details of the symptoms and deaths of gardeners who had inadvertently eaten monkshood tubers, having confused them with Jerusalem artichokes. The active poison, aconitin, is sometimes used as a narcotic and painkiller.

Anemone hupehensis,
Japanese anemone

Crassulaceae
THE STONECROP FAMILY

The succulent stonecrops and their kin are typically evergreen herbs or shrubs, but those from colder climates can also be deciduous, and there are a few annuals and aquatics.

Size

Almost half of the 1,380 species in this family belong to the large genera *Sedum* and *Crassula*. They have few uses other than as ornamentals, including the numerous stonecrops (*Sedum*), kalanchoes (*Kalanchoe*), jade plants (*Crassula*), and houseleeks (*Sempervivum*).

Sedum aizoon,
Aizoon stonecrop

Range

Though found almost worldwide (except in polar regions), these plants are common in arid areas and are often absent from tropical forests. They are especially diverse in southern Africa and Mexico, and rare in Australia and South America.

Origins

Succulent plants rarely preserve, so the history of this family is known only from a few recent pollen fossils.

Flowers

Stonecrop flowers have four or five sepals, which can be free or fused together to some degree. Similarly, their four or five petals can be free or fused, with the number of stamens either exceeding or sometimes equaling the number of petals. Individual flowers are usually radially symmetrical.

Fruit

The fruits are dry, as befits these desert plants, and are either follicles or, sometimes, capsules.

Echeveria secunda,
hens-and-chicks

Kalanchoe flammea,
Kalanchoe

Leaves

In *Crassulaceae* the leaves are always succulent and usually simple, though divided leaves do occur (*Kalanchoe pinnata*). They can be alternate, opposite, or whorled, with several genera forming tight rosettes (*Echeveria*, *Sempervivum*, *Aeonium*).

Propagation

Many succulents are easy to propagate, perhaps as a result of evolving in arid climates where mortality is always just around the corner. When the going gets tough, asexual reproduction allows new plants to form without the need for flowers or pollination. Gardeners can take advantage of this to produce free plants. Clump-forming stonecrops typically root whenever they touch the soil, so are readily divided. With rosette-forming plants, carefully remove small rosettes, or even individual leaves, and rest them on some compost; they'll quickly root. Stem cuttings work well for upright plants, while those with rhizomes (including many hardy perennials) can be split with a spade or knife. Some *Kalanchoe* even produce fully formed baby plants on their leaves, ready to pick off.

Hamamelidaceae

THE WITCH HAZEL FAMILY

Well known for the spectacular winter flowers of witch hazel, the family *Hamamelidaceae* contains some other early-flowering shrubs, such as winter hazel (*Corylopsis*), as well as the Persian ironwood (*Parrotia persica*)—a valued garden tree. Among the cultivated species there is a family preference for loamy, acidic soils.

Size

This medium-size family of trees and shrubs most notably includes the witch hazels (*Hamamelis*) and a few other ornamental species, such as *Corylopsis* and *Fothergilla* . There are about 29 genera and 95 species.

Range

Disconnected subfamilies exist across both hemispheres, in temperate and subtropical regions. There are five distinct centers: North and Central America, South and southwest Africa (as well as Madagascar), southwestern Asia, eastern and southeastern Asia, and northern Australia.

Fothergilla gardenii,
dwarf fothergilla

Origin

Originating during the Cretaceous, at the time of the dinosaurs, the *Hamamelidaceae* became very successful and widespread during the 65 million years that came before the relatively recent Quaternary glaciation (2.5 million years ago), which led to the extinction of many species and left isolated pockets of survivors. This is the reason why we see no family members in Europe.

Distylium racemosum,
isu tree

Flowers

It is difficult to generalize about the flowers in this family, since there is such a great variation. The flowers of the witch hazels, however, are incredibly distinctive with their four slender petals. Petals are absent entirely in the witchelders, which may come as a surprise to some gardeners. It is actually the many brushlike stamens that give the display.

Most typically, the flowers are carried in spikes or heads, and there is a family trait for early flowering, which makes some of these shrubs—particularly *Hamamelis* and *Corylopsis*—so useful in the late-winter garden.

Leaves

Most family members have fairly unremarkable, simple, elliptic leaves that are arranged alternately along the stems. Although unrelated to each other, the leaves of witch hazels closely resemble those of true hazels (*Corylus*).

USES FOR THIS FAMILY

As woody plants in the garden, the witch hazel family has much to offer, particularly those that are in a neutral to acidic soil. Witch hazels and *Corylopsis* are must-haves in any winter display, and *Fothergilla* not only produce excellent spring flowers, but also good leaf color in the fall.

Persian ironwood (*Parrotia persica*) is a good garden tree because of its extended season of interest. It has excellent fall color, attractively peeling bark, and, in late winter, its flowers on the branches are a welcome sight.

Parrotia persica,
Persian ironwood

Hamamelis virginiana,
common witch hazel

Paeoniaceae
THE PEONY FAMILY

This family includes many familiar shrubs and herbaceous perennials. They have a long history of cultivation in China, and there are also many species native to Europe and North America.

Size

There are 25 species currently described. It is what botanists would call a "monotypic" family; the *Paeoniaceae* has only one genus: the peonies (*Paeonia*). These are made up of herbaceous peonies—typically growing up to about knee or thigh height and dying back to ground level over winter—and deciduous "tree" peonies, which more closely resemble shrubs, although some moutan peonies are capable of reaching up to 65 feet.

In 1948 the Japanese nurseryman Toichi Itoh managed to cross tree and herbaceous peonies to produce a new category of peony called the Itoh or Intersectional hybrids. They share characteristics of both groups.

Range

Native only to the temperate regions of the northern hemisphere, there are three centers of distribution, chiefly southern and central Europe, Asia, and western parts of North America. In the wild, the tree peony species are confined to the northern Himalayas, but they are now cultivated worldwide.

Paeonia delavayi,
Delavay peony

USES FOR THIS FAMILY

Many gardeners are keen to have at least one peony in their garden. Grown as ornamentals as well as for their cut flowers, peonies are very alluring border plants, even though their flowering period can be brief. Due to extensive hybridization, there is a great range of flower forms and colors, from white to pink, red, deep scarlet, and purple, as well as a few bicolored forms. *Paeonia ludlowii* and its cultivars have yellow flowers.

Origin

The roots of the peony family exist in the order *Saxifragales*. While this order extends back to the fossil record of the Late Cretaceous (80 million years ago), fossil evidence from the Cenozoic Era suggests that peonies evolved from about 60 million years ago, when *Saxifragales* plants were more widespread.

Paeonia, once included in the buttercup family (*Ranunculaceae*), has important anatomical features in the flower that separate the genus from other families: persistent sepals, petals that are derived from sepals rather than stamens, and a fleshy nectar disc. The peony family is actually closer to the witch hazel family (*Hamamelidaceae*).

Flowers

It is the conspicuous and frequently beautiful spring and early summer flowers that peonies are known for, each one surrounded at the base by five green sepals and, occasionally, leaflike bracts. They are large, goblet-shaped, with five to ten petals. Both male and female parts are present. The many anthers are centrifugally arranged, and the two to five carpels are free and borne on a fleshy nectar disc; on pollination these ripen and dry into leathery follicles containing red seeds that mature to black.

Leaves

Peonies have distinctive leaves that push up through the soil in early spring on herbaceous species. They are usually deeply lobed and divided into several leaflets. Tree peony leaves tend to be larger and develop from buds on the woody stems.

Paeonia officinalis,
peony "Rubra Plena"

Flower in section

Single stamen

Seed in section

Ripe follicle showing seeds

Saxifragaceae
THE SAXIFRAGE FAMILY

In the past, this family included numerous woody plants, such as hydrangeas and currants, but now it is restricted to herbaceous plants. Often good yard plants, *Saxifragaceae* has produced numerous alpines (*Saxifraga*) and non-woody perennials (*Astilbe, Bergenia, Heuchera, Rodgersia*).

Size

The genus *Saxifraga* makes up more than half the approximately 625 species in this family, with many other genera comprising only one or a handful of species.

Range

Most *Saxifragaceae* occur in the northern hemisphere, often at high elevation or in arctic regions, with great diversity in eastern Asia and the Pacific Northwest. A few species can also be found farther south on tropical mountains.

Origins

Fossils from Eocene London Clay deposits (56–49 million years ago) represent the earliest record of *Saxifragaceae*, though, as with other herbaceous groups, this record is sparse.

Saxifraga flagellaris,
whipcord saxifrage

Flowers

The flowers appear in inflorescences of a variety of shapes, sometimes clothed in leafy bracts. Each bloom has four or five sepals (rarely three to ten) free or fused, and four or five petals (rarely three to ten), also free or fused. The petals may also be absent (as in *Rodgersia*, some *Astilbe*) or delicately lobed (*Mitella, Tellima*). Many *Saxifragaceae* (*Bergenia, Darmera*, many *Saxifraga*) offer a combination of charming foliage and attractive flowers; *Astilbe* are grown chiefly for their blooms, while *Heuchera* and its kin are predominantly foliage plants.

Tellima grandiflora,
fringe cups

Fringe petal

Flower in section with fringe petals attached.

Leaves

Great variety can be found in saxifrage leaves, ranging from the tiny, almost needlelike leaves of some alpine *Saxifraga*, to the huge, parasol-shaped foliage of *Darmera*. They are mostly alternate, although opposite in *Chrysosplenium* and some *Saxifraga*, and often clustered in a rosette. Simple leaves are common, though *Heuchera* leaves are maplelike, while those of *Astilbe* and *Rodgersia* are divided into leaflets. The various alpine saxifrages have a reduced number of leaves that can be finely divided, tongue- or spoon-shaped, forming neat hummocks or tight rosettes. Hairs are common on the foliage of many alpines as well as perennials, such as *Tellima* and *Tiarella*. Both *Tolmiea menziesii* and *Saxifraga stolonifera* produce plantlets, making them easy to multiply.

Bergenia purpurascens,
purple pigsqueak

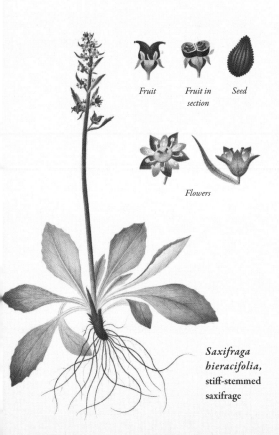

Fruit   *Fruit in section*   *Seed*

Flowers

Saxifraga hieracifolia, **stiff-stemmed saxifrage**

Euphorbiaceae
THE SPURGE FAMILY

The spurge family is large and complex. With huge variation among its members, it is difficult to attribute single features that unify the family. Even the biggest and best-known genus, *Euphorbia*, from which the family takes its name, shows great diversity—from woodland herb to desert succulent.

USES FOR THIS FAMILY

In the garden, a number of spring-flowering, perennial herbaceous species from the genus *Euphorbia* are popular. With its colorful flower bracts, the poinsettia (*Euphorbia pulcherrima*) is common both as a warm-temperate tree and as an indoor plant. In subtropical gardens, the shrubby crown-of-thorns (*E. milii*), with its small red flowers and very spiny stems, is planted widely. The tropical shrubs croton (*Codiaeum variegatum*) and chenille (*Acalypha hispida*) are both valued as houseplants in cool climates; the former for its distinctive leathery foliage and the latter for its long red cattail inflorescences. The castor oil plant (*Ricinus communis*) is sometimes grown in bedding schemes for its exotic palmate foliage.

Euphorbia pulcherrima,
poinsettia

Dalechampia spathulata,
winged beauty shrub

Size

With more than 6,500 species organized among 229 genera, this is a sizable family of flowering plants, most of which are herbaceous.

Range

While the majority of *Euphorbiaceae* species are restricted to the tropical zones, the genus *Euphorbia* in particular has footholds in some temperate areas, such as the southern United States, the Mediterranean, the Middle East, and South Africa.

Ricinus communis,
castor oil plant

*L–R: seed capsule,
individual castor seed
[not to scale]*

The richest concentration of tropical species is found in Indonesia and the Americas. A number of African euphorbias look just like cacti, which is the result of convergent evolution.

Origin

Molecular research by botanists at the UK's Royal Botanic Gardens Kew estimates the first plants of the *Malpighiales* order (of which *Euphorbiaceae* is a member) appeared about 91–88 million years ago. Fossil evidence suggests that the spurge family took form about 45 million years later.

Flowers

The individual flowers tend to be small and insignificant. Massed together in flower heads or as "false flowers" (pseudanthium), however, they can be quite showy—for example, the acid yellow-green flower clusters of marsh spurge (*Euphorbia palustris*) and the long cattails of chenille (*Acalypha hispida*).

The pseudanthium are formed by the close aggregation of the tiny flowers, along with colorful bracts and nectar glands to create a flowerlike effect. Examples include the winged beauty shrub (*Dalechampia spathulata*) with bright pink bracts; scarlet plume (*Euphorbia fulgens*) with bright red glands; and *E. corollata* with unusual white gland appendages.

Leaves

Spurge familiy leaves are mostly alternate along the stems, and simple in outline. Whenever they are compound they are always palmate (*Ricinus communis*) and never pinnate. The broken stems commonly produce a milky latex. In many cases this is poisonous and a skin irritant.

Acalypha hispida,
chenille

Salicaceae

THE WILLOW FAMILY

The best-known members of this family of trees and shrubs are willows (*Salix*) and poplars or aspens (*Populus*). Previously these genera were the only members of the family, but taxonomic changes have ushered in many tropical (and a few hardy) trees and shrubs.

Size

About 450 of the 1,200 species in *Salicaceae* are willows, and this number does not include hybrids.

Range

The willow family of old were largely restricted to the temperate northern hemisphere, but the addition of plants from the family *Flacourtiaceae* widened their range into the tropics and farther south. *Salix arctica* shares the record for the northernmost vascular plant, at 83°N.

Origins

A well-preserved willowlike fossil from Eocene deposits found in Utah and Colorado suggests that the plants of this family were well established at that time (53–48 million years ago).

Flowers

The small flowers on willows and poplars are clustered together in catkins, with male and female on separate plants. The flowers have no sepals or petals and each male flower has at least two stamens—sometimes many. In willows, the

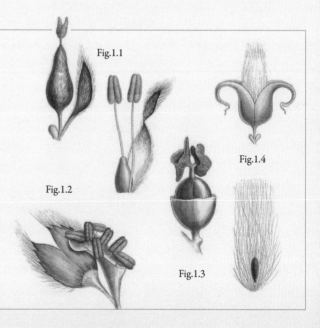

Fig.1.1. Female (left) and male (right) willow flowers lack petals and can be found in catkins on separate plants.

Fig.1.2. This male flower has several stamens and a hairy bract; when clustered together in catkins, the bract hairs give the appearance of pussy willow.

Fig.1.3. A female flower with bract removed, revealing the ovary and convoluted stigma; the bulge in the stalk is a nectary.

Fig.1.4. Willow fruits (above) are dry capsules that split open to release numerous small seeds with silky tassels (below), which aid in wind distribution.

Fig.1.1

Fig.1.2

Fig.1.3

Fig.1.4

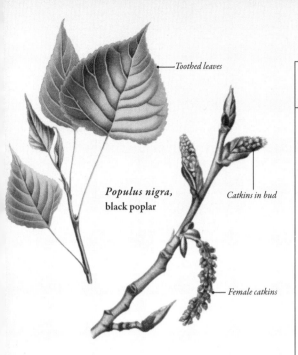

Toothed leaves

Populus nigra,
black poplar

Catkins in bud

Female catkins

paw-like catkins are upright and each tiny flower sits within a hairy bract, giving rise to the name "pussy willow." In poplars, the catkins are pendulous and the bracts can be hairy, smooth, or membranous. Other *Salicaceae* have flowers with sepals and petals in equal number, while some have no petals at all. The number of stamens ranges from one to numerous.

Idesia polycarpa,
igiri tree

USES FOR THIS FAMILY

Willow and poplar species readily swap genes with other species in their respective genera. The resulting offspring (hybrids) are often more vigorous than their parents, so lend themselves to soil stabilization, privacy screens, and biomass harvesting for energy production. Fully-grown willow or poplar trees are often too large for average yards, not to mention that their thirsty roots have been known to invade drains, causing subsidence and plumbing problems. Instead, opt for the especially fetching shrub *Salix fargesii.* Cut willow stems will root readily and, if planted in a row, create a cross between a fence and a hedge. They can also be woven into low hurdle fences or living wigwams for children. And it's not only willows that deserve garden space; the vanilla-scented flowers of *Azara microphylla* fill the winter air with fragrance.

Fruit

Fruits are often dry capsules (seeds with silky hairs in willow and poplar), though berries and fleshy capsules also occur.

Leaves

The willows, poplars, and aspens are deciduous, with the latter developing often excellent coloration in the fall. Other temperate genera, such as *Poliothyrsis, Carrierea,* and *Idesia* are also deciduous, while *Azara* and most other tropical species are evergreen. The leaves are alternate (rarely opposite) with distinct petioles, and some species (including most willows) have prominent stipules.

Violaceae
THE VIOLET FAMILY

The violets and pansies constitute the largest part of this modestly sized family, being small herbaceous plants with pretty flowers. Other family members are much more obscure because they do not share the same ornamental qualities.

Size

Viola is the main genus, with about 700 different species. Some are annual or short-lived, such as heart's ease (*Viola tricolor*), a pretty herb that creeps over short grass or wasteland, while the great majority are nonwoody perennials, such as *Viola riviniana* (dog violet)—no relation to the dog's tooth violet (*Erythronium*), which is of the lily family (see pages 78–79). A further 300 species make up the rest of the family. Note that African violets (*Saintpaulia*) are not related to this family.

Flower

Fruit

Range

While true violets (*Viola*) are found predominantly in temperate regions of the northern hemisphere, representatives from across the family can be found worldwide. In the tropics, *Violaceae* family members tend to be restricted to areas of higher altitude.

Origin

Violaceae is not represented in the fossil record, making its origins difficult to confirm with certainty. Its placement in the order *Malpighiales* dates the family's possible origin to the end of the Cretaceous (80–70 million years ago), with the evolution of true violets taking place much later.

Flowers

With the exception of the true violets (*Viola*), all flowers of the *Violaceae* are regular, with five sepals, five petals, and five stamens. They are carried in racemes or panicles, or individually from the leaf axils.

Melicytus crassifolius,
thick-leaved mahoe
Tree violets are tough shrubs that are sometimes grown as ornamentals in their native Australia and New Zealand.

Viola tricolor,
wild pansy

Viola lutea,
mountain pansy

Viola flowers have unequal petals with the lowermost pair, often the largest, forming a prominent spur. Petal color varies with shades of blue, yellow, white, and cream, some bi- or multicolored. Flowering is most profuse in spring and early summer, but often extends into other times of the year.

Pollination

Scent and linear markings on the colorful petals guide bees to the flowers, and a nectar reward in the spur encourages them to advance further, causing the bee's body to touch the female stigma. The insect also touches the anthers, which shower pollen on its back. Flying from flower to flower, a single bee can cross-pollinate hundreds of flowers in a day.

USES FOR THIS FAMILY

Violets and pansies are extremely popular among gardeners for seasonal bedding or container displays. British horticulturists in 1979 developed pansies that are able to bloom in the short days of winter, which has changed winter bedding displays forever. Smaller species, such as sweet violet (*Viola odorata*) and horned violet (*V. cornuta*), are more subtle but no less beautiful. Some violets, such as the common blue violet (*V. sororia*), can naturalize in shady lawns, which might either be welcomed or considered a nuisance.

Viola odorata,
sweet violet

Viola cornuta,
horned violet

Fabaceae

THE LEGUME FAMILY

It is perhaps in the vegetable garden that this family proves its worth; the characteristic pods of annual peas and beans are especially significant. Commercially important legumes include soy; lentils; peanuts; and forage crops, such as alfalfa and clover. But worldwide, most legumes are woody trees and shrubs, including important forest and timber trees such as acacia, rosewood (*Dalbergia*), and locust (*Gleditsia*, *Robinia*).

Size

Though it's the third-largest plant family, with more than 19,500 species, many are tropical and of limited use in temperate yards. Important exceptions include trees such as laburnum and redbud (*Cercis*), climbing wisteria and sweet peas, shrubby wattles (*Acacia*) and broom (*Cytisus*, *Genista*, *Argyrocytisus*), and herbaceous lupines and false indigo (*Baptisia*).

Single flower

Leaflike phyllodes

Pisum sativum,
garden pea

Acacia acinacea,
gold-dust wattle

Range

This family can be found worldwide, though not in Antarctica or the northern polar regions. However, it is most diverse in the tropics and is an important component of wet and dry tropical forests. The dry woodlands of sub-Saharan Africa, for example, would be almost nonexistent without acacias and their relatives.

Origins

Studies of fossils and DNA suggest the legumes originated in the Paleogene, with the earliest fossil dated to around 56 million years ago. A period of diversification occurred in the Eocene, when several recognizable legume groups developed.

Lathyrus odoratus,
sweet pea

USES FOR THIS FAMILY

The flowers of many hardy legumes have evolved to be pollinated by bees, so adding these to your yard will help to provide food for these important pollinators. A good place to start is the lawn and allow legumes, such as clover (*Trifolium*) and trefoil (*Lotus*), to grow and flower among grass. If you must have a closely cropped sward then leave a few edges to grow wild. Scarlet string beans (*Phaseolus coccineus*) and sweet peas are also bee magnets; grow them up your boundary fence or on obelisks in your herbaceous borders.

Flowers

Many different flower types can be found in *Fabaceae*, but for gardeners, it is the pea-style that is most familiar. These butterfly-like blooms have five sepals, usually fused partially into a tube, plus five petals differentiated into three structures: standard, wings, and keel. The uppermost petal forms the showy standard (or banner), two laterals are wings providing a landing spot for insects, while the lowermost two are partially fused together to create the keel. Ten stamens, sometimes fused together, are concealed within the keel, emerging when pollinators land on the flower. To this basic form we can add flowers with small petals and showy brushes of stamens (*Acacia, Albizia*), petals that all look alike (*Senna, Delonix*), or those flowers largely with the petals fused together into a tube (clover).

Trifolium reflexum,
buffalo clover

Fruit

The fruit of legumes, itself called a legume, is perhaps this family's most identifiable characteristic. Often described as a pod, this dry, elongated fruit splits along both seams to reveal a single row of seeds. If you've ever shelled peas or beans then you'll be very familiar with a typical legume fruit. Several other fruit types can be found in the family, but pods are by far the most common. The pods can be attractive in their own right, as in the purple pods of hyacinth bean (*Lablab purpureus*). To distribute their seeds, some legume pods pop, as with broom and lupine, while others simply hang open and drop seeds. Those of carob (*Ceratonia*), tamarind (*Tamarindus*), and honey locust (*Gleditsia*) contain sweet pulp that attracts hungry animals.

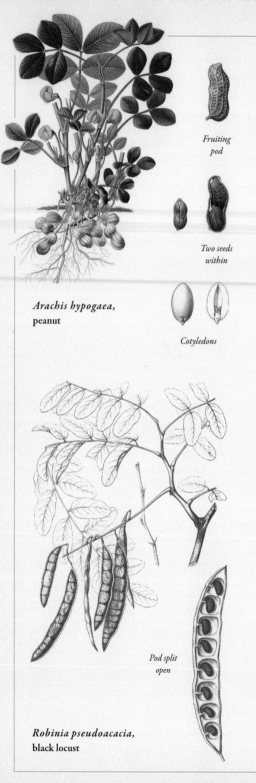

Fruiting
pod

Two seeds
within

Arachis hypogaea,
peanut

Cotyledons

Robinia pseudoacacia,
black locust

Pod split
open

Lens culinaris,
lentil

Lablab purpureus,
hyacinth bean

Leaves

Most legumes have compound leaves with their leaflets in a feather-like (pinnate) arrangement. In some cases, leaflets are in a handlike (palmate) arrangement (lupine, clover), or are divided into smaller leaflets (*Acacia, Albizia*). A few species, such as redbud (*Cercis*), have entire leaves. The arrangement is alternate, and stipules are usually present, often appearing leaflike, as with peas (*Pisum*), or developing into spines, like locust (*Robinia*). In climbing sweet peas (*Lathyrus odoratus*), the terminal leaflet is replaced by a tendril. Many Australian acacias appear to have simple leaves, but these structures are flattened petioles known as phyllodes. Their seedlings typically exhibit the more usual compound leaves. A conspicuous characteristic of many legumes is that their leaves fold up at night; for example, the leaves of the sensitive plant (*Mimosa pudica*) fold when touched by hungry bugs or curious kids.

Nitrogen fixing

All plants need to secure supplies of three basic nutrients—nitrogen (N), phosphorus (P), and potassium (K)—but nitrogen can be difficult to absorb, despite making up about 78 percent of the atmosphere. Plants cannot take it directly from the air and it rarely reacts with other elements to form soluble compounds that the roots can absorb. Many legumes get around this problem by forming nodules on their roots. These become infected with bacteria that can transform atmospheric nitrogen into fixed nitrogen, which the plants then absorb. Not only does this mean that legumes can thrive on nutrient-poor soils, but also they leave behind in the soil nitrogen for other plants to use. Gardeners can take advantage of this by using legumes in crop rotation, or as green manures or cover crops. Once the legumes are harvested, nutrient-hungry crops, such as corn, can be planted and will benefit from the residual nitrogen compounds in the soil.

Delonix regia,
flame tree

Moraceae
THE MULBERRY FAMILY

This is a family largely made up of trees, shrubs, and large tropical climbers called lianas. The flowers are usually tiny and form aggregate fruit, such as the well-known edible figs and mulberries. Shared traits include milky sap, inconspicuous flowers, and compound fruits.

Size

The *Moraceae* has about 1,150 species across 38 genera. The common fig (*Ficus carica*) is just one example from the most significant genus, *Ficus*, which has about 850 representatives across the world's tropical zones. These include shrubs, vines, epiphytes, trees, and grotesque stranglers that enclose and envelop the trees that they grow over. The genus *Morus* contains about a dozen species of mulberry tree, all deciduous, that grow well in temperate zones.

Some authorities include the genera *Cannabis* (hemp) and *Humulus* (hops) in this family, but they differ with their five-parted flowers, dry fruit (achene), and herbaceous stems. They belong to their own family: *Cannabaceae*.

Morus nigra,
black mulberry

Range

Moraceae family members are distributed worldwide across the tropics and subtropics. There are a few well-known representatives of temperate regions, such as the black mulberry (*Morus nigra*) and the common fig (*Ficus carica*). Strangler figs (*F. aurea*) are a vital component of many rain forest ecosystems. Iroko (*Milicia excelsa*) is a valued hardwood tree from tropical Africa.

***Broussonetia papyrifera*,**
paper mulberry
A curious shrub with no two leaves quite the same shape. The aggregate female fruits are borne in spherical heads about 1 inch wide.

Origin

Analysis of DNA sequences allows evolutionary biologists to estimate that *Moraceae* ancestry dates back about 80 million years. The main evolution of existing species, however, probably took place between 40 and 20 million years ago. There are close links with the nettle family (*Urticaceae*).

Flowers

Moraceae flowers are unisexual, which means they are either male or female. Flowers of both sexes may be on the same plant or on separate plants. Individually they are very small with four perianth segments. The flowers are aggregated together on flattened or bowl-like receptacles, or in catkins.

Fruit

Fruits are highly variable, but frequently fleshy, and occasionally edible. The fleshy parts are not derived from the flower's ovary, but from the receptacle on which the flower sits.

Artocarpus altilis, **breadfruit**
The large, rough-surfaced fruit of most varieties are seedless, making them easier to eat.

Pollination of figs

Fig species are characterized by their half-closed, bowl-shaped receptacle, which encloses the tiny flowers inside. They are pollinated by female fig wasps, which force their way into the receptacle and pollinate the flowers within. In return for this service, about one-third of the developing seeds are sacrificed as food for the hatching larvae.

Ficus carica,
common fig
The distinctive fruit of the garden fig is a complex structure, an aggregation of many smaller, one-seeded fruits. They ripen from green to brown.

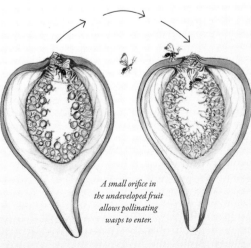

A small orifice in the undeveloped fruit allows pollinating wasps to enter.

Rosaceae
THE ROSE FAMILY

For gardeners, this is perhaps the most significant of all plant families. Encompassing trees, shrubs, herbaceous perennials, climbers, annuals, and alpines, there is a rose relative for every occasion. The commercial significance of the family is also huge because it includes apples, almonds, pears, plums, peaches, cherries, strawberries, raspberries, and several other fruit-bearing species.

Size

There are more than 3,000 members of *Rosaceae*. While the fruit-bearing species are especially well known, this family also includes decorative trees and shrubs such as mountain ash (*Sorbus*), hawthorn (*Crataegus*), firethorn (*Pyracantha*), cotoneaster, and, of course, rose (*Rosa*). Perennial gardens also rely on many *Rosaceae*, for example lady's mantle (*Alchemilla*), avens (*Geum*), cinquefoil (*Potentilla*), and meadowsweet (*Filipendula*).

Rubus spectabilis, salmonberry

Prunus domestica, plum

Range

Found on every continent but Antarctica, the rose family is especially diverse in the northern hemisphere. It is much less common in arid or humid tropical areas and is an important component of temperate forests.

Origins

The very earliest fossils of *Rosaceae* date back to around 100 million years ago. Their flowers are relatively unspecialized, allowing a wide range of insects to visit them, unlike in some more advanced families that have evolved complex, often highly specific relationships with their pollinators.

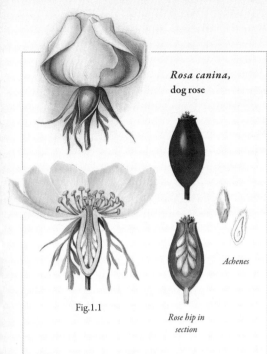

Rosa canina,
dog rose

Fig.1.1

Achenes

Rose hip in
section

Fig.1.1. In roses, several carpels are enclosed within a cuplike structure called a hypanthium. Once pollinated, the hypanthium becomes the fleshy hip, but the real fruits are the dry achenes inside.

Fig.1.2. Cherry blossoms also have a cuplike hypanthium, but with a single carpel inside, which develops into a single fleshy fruit containing a single seed.

Fig.1.3. With petals and hypanthium removed and the ovary in section, the single seed developing inside the ovary is apparent.

Sanguisorba officinalis,
great burnet

Flowers

The simple bloom of dog rose (*Rosa canina*) is a good example of the basic flower structure in the *Rosaceae*. Five green sepals protect the bud, which opens to reveal five colorful petals and a dense cluster of stamens. In many roses, flowering cherries (*Prunus*) and others, breeders have increased the number of petals to form lush double blooms, where the stamens are almost invisible or completely absent. Other family members, such as lady's mantle and burnet (*Sanguisorba*), have no petals at all. Fragrance is not uncommon in *Rosaceae* and rose blooms are grown commercially to extract oil for perfumery. However, not all species have scented flowers and some smell unpleasant; thankfully, the meaty scent of mountain ash (*Sorbus*) blooms is short-lived.

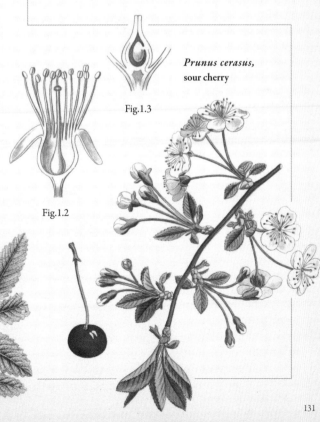

Prunus cerasus,
sour cherry

Fig.1.3

Fig.1.2

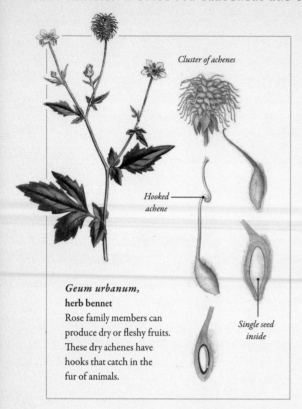

Cluster of achenes

Hooked achene

Single seed inside

Geum urbanum,
herb bennet
Rose family members can produce dry or fleshy fruits. These dry achenes have hooks that catch in the fur of animals.

Fruit

Across this large family, the fruits that develop are extremely variable. Plants with dry fruits can be ornamental; for example, avens (*Geum, Dryas*) and piri-piri (*Acaena*). Fleshy fruits are especially common and important to gardeners and farmers. Apples, plums, raspberries, and strawberries represent the four primary types of fleshy fruit. The main difference between these is in how the flesh develops from the flower. Apples contain seeds within a leathery core, surrounded by flesh that forms from the receptacle.

Prunus armeniaca,
apricot

In plums, the seeds are enclosed in a hard shell or pit, and the flesh develops from a single carpel. Raspberries are composed of numerous tiny fruits, structurally similar to plums, but clustered together. Finally, strawberries aren't fruits at all; the fleshy part is formed from the receptacle, while the pips are the real fruit, formed from the carpels.

Leaves

While there's considerable variation in the leaves of *Rosaceae*, they usually alternate along the stems in trees and shrubs, or cluster at the base of herbaceous plants. They can be simple and undivided, or divided into feather-like (pinnate) or handlike (palmate) arrangements. They also usually have a pair of small leaflike tufts at the base of each leaf stalk (stipules). Most species are deciduous, often providing pleasant color in the fall, as in chokeberry (*Aronia*) and rowan. There are several evergreens, including cherry laurel (*Prunus laurocerasus*), loquat (*Eriobotrya japonica*), firethorn (*Pyracantha*), and many species of *Cotoneaster*.

Malus domestica,
apple

Sorbus aucuparia,
mountain ash

Pollination

In most plants, reproduction involves the transfer of pollen from the stamens of one flower to the stigmas of another. This facilitates the transfer of genes and results in seeds and seedlings that are genetically distinct from their parents. When two different species cross-pollinate, either the resulting seed is infertile or hybrids are produced, which usually appear intermediate between the parents. Keep this in mind when collecting seeds from garden-worthy plants; the seedlings produced may not share the positive qualities of the fruiting parent. Such hybrids are typically infertile and cannot produce seed of their own, but several family members are exceptions, including some lady's mantle, hawthorn, blackberry (*Rubus*), and mountain ashes. Their hybrids can produce seeds without any genetic transfer and the resulting seedlings are genetically identical to the seed parent. In these cases, it is safe to sow the seeds, knowing you'll get exactly what you want.

USES FOR THIS FAMILY

Without the *Rosaceae*, the orchards, gardens, and woodlands of the United States would be strangely empty. The family provides many other kinds of useful plants. For an attractive hedge, it's hard to beat hawthorn, cherry laurel, *Photinia × fraseri* "Red Robin," or, in coastal areas, the rugose rose (*Rosa rugosa*). Ground-cover plants such as *Waldsteinia ternata*, *Acaena caesiiglauca*, and *Rubus pentalobus* reduce the burden of weeding, while climbing and rambling roses are readily trained over arches and bowers. Perhaps the most useful attribute of this family is their value to wildlife; flowers that attract a range of insects and tasty fruits ensure that any garden filled with *Rosaceae* will also be home to abundant wildlife.

Rosa multiflora,
rambling rose

Begoniaceae
THE BEGONIA FAMILY

Most gardeners will be familiar with begonias, perhaps as blowsy bedding in a hanging basket or a much-loved pot plant on a windowsill. These herbaceous plants or shrubs can have fibrous roots, rhizomes, or tubers, while a few species are woody climbers or annuals.

Size

There are around 1,400 species in this family, though all but one are in *Begonia*, making it the sixth largest flowering plant genus. The single exception is *Hillebrandia sandwicensis*, a native of the Hawaiian islands.

Begonia "Prestoniensis,"
Preston begonia

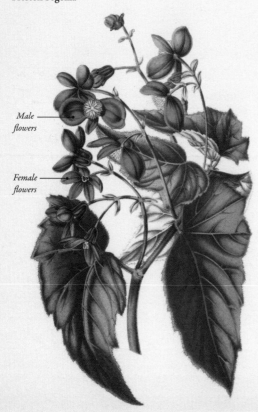

Male flowers

Female flowers

Range

Begoniaceae are restricted to tropical and subtropical areas of Asia, Africa, and the Americas, though strangely absent from Australia. Begonias are especially common growing on branches of tropical trees or on the rain forest floor.

Origins

The soft stems and leaves of *Begonia* are poor candidates for fossilization. A recently discovered fossil exists from the Pliocene (5–3 million years ago), but other analyses indicate the family probably dates to the Eocene or Oligocene (45–30 million years ago).

Flowers

In *Begonia* the genders are separated, with male and female flowers on the same plant or, rarely, on separate plants. Flowers emerge from leaf axils and can be clustered or solitary. Sepals and petals appear similar (tepals) and are usually white or pink, though less often red, orange, or yellow. Male flowers have two pairs of tepals, with one pair larger than the other, and a cluster of yellow stamens. Female flowers have five tepals, below which is a prominent three-sided, winged ovary; the flowers also appear to have yellow stamens; however, these are in fact stigmas. The number of tepals can vary beyond these basic forms, with many cultivated begonias having numerous additional tepals. The male and female blooms of the Hawaiian *Hillebrandia* have five sepals and five petals.

Leaves

Begonia coriacea,
leatherleaf begonia

Perhaps the most distinctive feature of begonia leaves is that they are not symmetrical, and often resemble the shape of a human ear. The leaves are alternate—though on the rare occasion opposite—and succulent, usually with petioles and prominent stipules. Most species have entire leaves, though they can be compound with pinnate (*Begonia bipinnatifida*) or palmate (*B. luxurians*) leaflet arrangements. A huge diversity of color and pattern exists in begonia foliage, which is one of their most important horticultural attributes.

Begonia diadema,
coronet begonia

USES FOR THIS FAMILY

With their many different growth forms, begonias have a variety of uses. Unlike most other summer bedding plants, tuberous and wax begonias will tolerate some shade, so choose them for north-facing borders and containers. Plant the tender *Begonia luxurians* for a touch of the exotic outside during summer; its giant leaves will dazzle your neighbors. By contrast, herbaceous perennial *B. grandis* can be left outside year-round and will survive as subterranean tubers during the winter. Indoors, choose one of the many excellent foliage begonias, such as hairy-leaved *B. sizemoreae* or silver *B. imperialis*, or create a terrarium in an old glass jar using the delicate *B. rajah*.

Begonia grandis,
hardy begonia

Cucurbitaceae
THE SQUASH FAMILY

Collectively known also as the cucurbits, the squash family is renowned for its swollen fruits: pumpkins, squashes, zucchini, and melons. A few species make woody lianas (large jungle climbers) and one, the cucumber tree (*Dendrosicyos socotrana*), native to the Arabian island Socotra, is known for its extraordinary flora and makes a very unusual, baobab-like tree.

Size

With 122 genera and about 940 species, these are typically climbing annual, herbaceous plants that are destroyed by frost.

Range

This family can be found in all tropical and subtropical regions of the world, particularly in the South American rain forests and African bushlands. A few species are adapted to drier semidesert habitats, where they can be an important source of food and water to indigenous people. Many cultivated varieties are grown as annual food plants in temperate regions around the world.

Cucumis melo,
melon

Momordica balsamina,
balsam apple

Origin

A number of fossil records exist for the squash family, but the earliest to date comes from a prehistoric species called *Cucurbitaciphyllum lobatum*, which is from the Palaeocene (61–56 million years ago).

There is some mystery surrounding the exact evolutionary origin of the cucurbits, since they bear very little relation to other plant families with which they are sometimes grouped, such as the *Begoniaceae*. The specialized anatomy and biochemistry of the cucurbits suggests a highly evolved nature.

Leaves

Common features of the *Cucurbitaceae* are their simple, palmately veined leaves arranged alternately along the five-angled, hairy stems. Sometimes the leaves are broad and palmately lobed.

On many species a single tendril grows from the base of each leaf, and its tip will curl around any nearby object that might provide support. The rest of the tendril then spirals into a spring, pulling the stem close to its support.

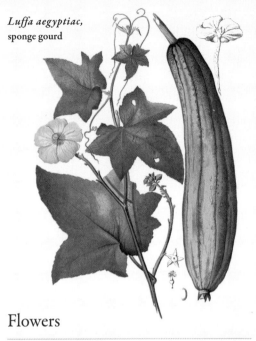

Luffa aegyptiac, sponge gourd

Cucumis sativus, cucumber

Female flower in section

Male flower in section

The female flower has a swollen base beneath the flower, which is the unfertilized ovary in an inferior position.

Flowers

The flowers are either male or female (unisexual), sometimes on the same plant, sometimes on separate plants. The female flowers clearly show what botanists refer to as an inferior ovary, where the petals and sepals are positioned above the ovary. As the ovary swells, forming the fruit, this arrangement is plain to see. The yellow or white petals are frequently large and more or less united at the base.

Fruits

Cucurbit fruits can be strange and beautiful, from swan-neck gourds to warty-looking pumpkins in a great many colors. These fruits are actually berries, defined as fleshy fruits without a stony or woody layer, and containing many seeds.

Often fruits have a firm wall (such as watermelons and pumpkins); at other times they are dry and leathery. *Luffa cylindrica* is the source of the loofah sponge, which is basically a dried skeleton of the plant's fruit.

Betulaceae

The birch family

The trees and shrubs of *Betulaceae*, which include alder (*Alnus*), birch (*Betula*), hornbeam (*Carpinus*), and hazel (*Corylus*) are grown for their pleasing habit, neatly veined leaves and often colorful peeling bark. The edible seeds of hazel and filbert are grown commercially.

Size

This is a small family of around 140 species, most of which fall into the four aforementioned genera, of which *Betula* is the largest. There are two additional genera, making in total six; the hop hornbeams (*Ostrya*) and hazel-hornbeams (*Ostryopsis*) complete the sextet.

Range

The trees of the birch family are common in temperate Europe, Asia, and North America. They reach into the tropics only at high elevation. The black alder (*Alnus glutinosa*) intrudes into North Africa, while Andean alder (*A. acuminata*) reaches into northern Argentina.

Origins

Betulaceae present a rich array of fossils, indicating the family originated in the Palaeocene (65 million years ago), with recognizable genera appearing around 45 million years ago.

Flowers

All *Betulaceae* produce catkins, which are not known for their showy flowers, but the way that they jostle in the wind gives them some charm. The male and female flowers are produced in

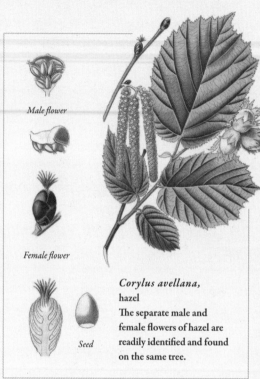

Male flower

Female flower

Seed

Corylus avellana, hazel
The separate male and female flowers of hazel are readily identified and found on the same tree.

separate catkins on the same tree. Male catkins dangle, releasing pollen into the wind. Female catkins can be upright (*Corylus, Ostryopsis*) or pendulous with prominent bracts (*Carpinus, Ostrya*). Catkins are made up of numerous scalelike bracts and each holds one to three flowers. Individual blooms are very small with either few or no sepals or petals. In alder, the female catkin resembles a small cone, while in hazel, it is hugely reduced and barely visible except for the emerging bright-red stigmas.

Fruit

All *Betulaceae* have dry, single-seeded fruits enclosed within the catkins. Birch catkins disintegrate, releasing tiny winged fruits, while in hornbeam the bracts form wings to help disperse the fruits. Hazelnuts form within elaborate clusters of leafy or sometimes spiny bracts.

Leaves

The basic *Betulaceae* leaf is alternate and simple with toothed margins and small, deciduous stipules. Foliage is most often deciduous, with some species producing attractive, though short-lived fall coloration. A few species are evergreen, or at least semievergreen, such as *Alnus jorullensis*. The often striking leaf venation gives a neat appearance to hornbeams and many others.

Female flower

Male flower

Carpinus betulus,
European hornbeam

Betula pendula,
silver birch

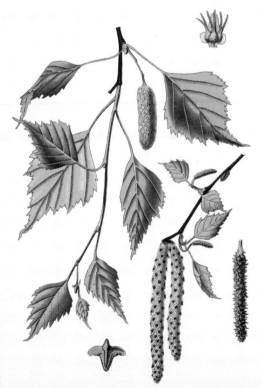

USES FOR THIS FAMILY

The peeling bark of some birches is showy and often colorful, earning these trees a place in many mid-size yards. Best known is Himalayan birch (*Betula utilis* var. *jacquemontii*), whose ghostly white stems transform the winter landscape, especially when planted in a group. In contrast, Chinese red birch (*B. albosinensis*) has delicate pink bark, which peels off in sheets. And birches are not the only beautiful trees; in late summer the female catkins drip from the branches of the monkey-tail hornbeam (*Carpinus fangiana*), contrasting with this tree's large, neatly veined leaves and making an arresting sight.

Fagaceae
THE OAK FAMILY

There are many charismatic trees in *Fagaceae*, including oak (*Quercus*), beech (*Fagus*), and chestnut (*Castanea*). They are valuable sources of timber and other forest products, such as edible chestnuts and cork. They are also important landscape trees, while several species form smaller shrubs.

Size

Of the 970 or more species in this family, around half (430) are oaks, and many make useful garden or landscape trees. While the dainty, holly-like *Quercus monimotricha* rarely reaches 6.5 feet, the lofty red oak (*Q. rubra*) often reaches twenty times that height. It is best suited to parks and open landscapes.

**Fagus
sylvatica,**
European beech

Lithocarpus daphnoideus,
daphne stone oak

Range

Fagaceae are common in temperate and tropical regions of the northern hemisphere. Oak, beech, and chestnut sweep across the northern continents, while the remaining genera are restricted to Asia (*Castanopsis*, *Lithocarpus*) or North America (*Chrysolepis*, *Notholithocarpus*). The obscure *Trigonobalanus* occurs in Colombia and southeast Asia.

Origins

As with *Betulaceae*, the fossil record is rich in *Fagaceae*, indicating origins in the Late Cretaceous, about 82 million years ago.

Flowers

All oak family members have separate male and female flowers, usually on the same tree and sometimes in the same inflorescence. With most genera, male flowers appear on fingerlike spikes; in other cases they are on pendulous catkins (oaks) or heads (beech). Female flowers can be in groups of two to three, in spikes or solitary. The sepals and petals are inconspicuous, but female flowers are surrounded by several overlapping bracts, which develop into a structure known as a cupule that contains the fruit.

Fruit

In oaks and the closely realted *Lithocarpus* and *Notholithocarpus*, each female flower produces a single round acorn with a scaly cupule. In beech and the remaining genera, each flower forms one or more angled nuts, enclosed within spiny cupules.

Leaves

Fagaceae can be either deciduous or evergreen. The deciduous types produce some of the most exciting displays of fall color, while evergreens have a glossy sheen and, in some cases, contrasting fuzzy coatings on the leaf undersides.

Female flower in section

Male flowers

Castanea sativa, European chestnut
Chestnuts form within a spiny cupule, which splits open at maturity.

Quercus montana, chestnut oak

Quercus velutina, black oak

Leaves are usually alternate and simple, but some oaks show a complex array of lobes. Stipules are present, though these are best looked for on new growth. Leaf margins are entire or serrate and petioles are usually swollen at the base.

Masting

Traditionally, in the fall, many pig farmers moved their animals into deciduous woodland to fatten them up on acorns and beech nuts, known collectively as "mast." Some years the trees would produce a bumper mast crop, which not only helped the farmers but also improved survival for tree seeds. This synchronization of mass fruiting is called masting and is thought to overwhelm seed predators, such as squirrels and jays (as well as their domestic cousins), ensuring some seeds are left to germinate. The exact process that allows trees across a wide area to synchronize the size of their seed crop is still unknown.

Juglandaceae
THE WALNUT FAMILY

Best known for the walnuts and hickories, this family contains many other large hardwood trees that are valued for their timber as well as their nuts. Their impressive stature is a quality they share with closely related families *Fagaceae* (beeches and oaks) and *Betulaceae* (birches and hornbeams).

Size

This is a small family of just 61 species of hardwood deciduous trees, shared across eight genera. The family's best-known members are the walnuts (*Juglans*) and the hickories (*Carya*). Further subdivision of the family into two subfamilies is possible. Subfamily *Juglandoideae* contains just two genera: *Juglans* and *Carya*. Subfamily *Oreomunneoideae* contains lesser-known genera, such as *Pterocarya* (wingnut), *Engelhardtia*, *Oreomunnea*, and *Platycarya*.

Engelhardtia spicata,
Lesch ex Blume.

USES FOR THIS FAMILY

The family is best known for its edible walnuts (*Juglans regia*), pecan nuts (*Carya pecan, C. illinoensis*), and hickory nuts (*C. ovata*). They are also valued landscape trees, some of which have good fall foliage, but their size restricts their use to large gardens and parks.

Carya illinoinensis,
pecan

Range

The subfamily *Juglandoideae* is confined to temperate and subtropical areas predominantly in the northern hemisphere, with some American species extending south into the Andes. The *Oreomunneoideae* subfamily has a more tropical bearing, with representatives in Central America, and south to southeastern Asia and China.

Origin

The fossil record suggests that all major lineages of the *Fagales* order (of which the *Juglandaceae* is a member) are rooted in the Late Cretaceous (about 97 million years ago). Specific fossils from the walnut family begin to appear after about 66 million years ago, in the Palaeogene.

Flowers

The unisexual flowers are borne together on the same plant, with the male catkins hanging from last year's stems and the small female spikes on the new spring growth. The individual flowers are small and wind-pollinated. When massed together, the display can sometimes be quite pretty, especially when the winged fruits begin to form, as with the wingnuts (*Pterocarya*). It can sometimes be hard to examine the outer part of the flower because it is often tiny or shrivels and falls early.

Fruit

Subfamilies *Juglandoideae* and *Oreomunneoideae* are distinguished by their fruits, which in the first group are drupes and in the second winged nuts. The fruits of wingnuts do not have a fleshy coating but instead have a single dry nut with two wings on opposite sides. The wheel wingnut (*Cyclocarya paliurus*) has a wing that encircles the nut. Walnut and hickory fruits are a type of stone fruit; they have a fleshy outer layer that encloses a woody "stone," which is typically hard to crack open. The stone contains the seeds, which in this case are the popular pecans, hickories, and walnuts. The stones of *Juglans* have a sculptured surface, whereas those of *Carya* are smooth.

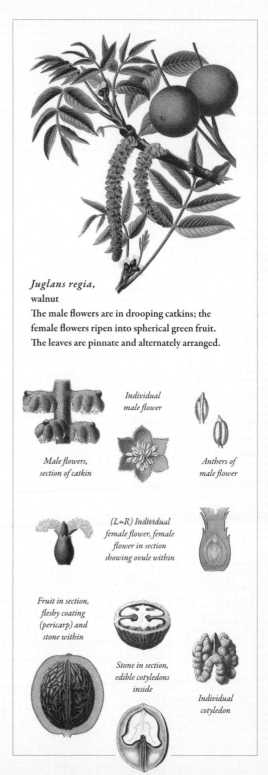

Juglans regia,
walnut
The male flowers are in drooping catkins; the female flowers ripen into spherical green fruit. The leaves are pinnate and alternately arranged.

Individual male flower

Male flowers, section of catkin

Anthers of male flower

(L–R) Individual female flower, female flower in section showing ovule within

Fruit in section, fleshy coating (pericarp) and stone within

Stone in section, edible cotyledons inside

Individual cotyledon

Geraniaceae
THE CRANESBILL FAMILY

This family of herbaceous perennials, small shrubs, and the occasional annual includes the horticulturally significant cranesbills (*Geranium*), storksbills (*Erodium*), and geraniums (*Pelargonium*). Many curious succulent plants can be found in *Pelargonium* and *Monsonia*.

Geranium pratense,
meadow cranesbill

Size

Most of the 650 species of *Geraniaceae* fall into either *Geranium* or *Pelargonium*. The two have long been confused and the first *Pelargonium* to reach Europe from Africa was initially placed in the European-native *Geranium*. The two were later separated, but not before the common name "geranium" had been applied widely to both.

Range

The family occurs on every continent except Antarctica, though they are most common in temperate and subtropical zones. In the tropics they are typically found at high elevations.

Geranium robertianum,
herb robert

Origins

A pollen fossil around 28 million years old (from the Oligocene) has been identified as *Geraniaceae*. This relatively young family probably dates to the boundary between the Eocene and Oligocene, around 33 million years ago.

Flowers

The main attraction of geraniums is their colorful flowers, which are mostly pink, red, purple, or white. Most are radially symmetrical, except in *Pelargonium* where they are slightly bilateral. Each bloom has five sepals that are free or slightly fused

The herbaceous storksbills and cranesbills are useful border perennials whose flowers are much beloved by bees. Some of the daintier *Erodium* thrive in the well-drained soils of rockeries, wall crevices, and between pavers. *Pelargonium* also provides many valuable garden plants, especially the zonal, regal, and trailing ivy-leaved forms that are so useful in bedding schemes and hanging baskets. Unlike many bedding plants, they are drought tolerant, reducing the need for endless and wasteful watering. The many scented-leaf geraniums make excellent container plants and are also thought to ward off some insect pests, so are good to grow among vegetables as well as in pots near the kitchen door.

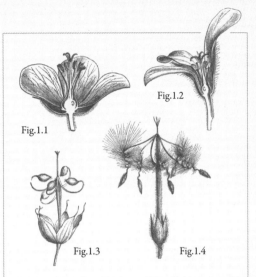

Fig.1.1

Fig.1.2

Fig.1.3

Fig.1.4

Fig.1.1. *Geranium* flower in section; it is radially symmetrical, with all petals of equal size.

Fig.1.2. *Pelargonium* flower in section; it is bilaterally symmetrical, with slightly bigger petals on one side.

Fig.1.3. *Geranium* fruits (or schizocarps) have five sections, which split rapidly when dry, catapulting their seeds far away from the parent plant.

Fig.1.4. *Pelargonium* schizocarps also split open, but their seeds typically bear feathery parachutes allowing them to be dispersed by the wind.

at the base, and five free petals. In many species the petals have noticeable lines to guide pollinators, while in some *Erodium* and *Pelargonium* the upper two petals have colorful mottling. Most genera have five, ten, or fifteen stamens, though in *Pelargonium* there are two to seven.

Leaves

Many *Pelargonium* are known for their fragrant foliage, with scents reminiscent of roses, mint, citrus, or chocolate. On the other hand, the leaves of herb robert (*Geranium robertianum*)— also known as "stinky Bob" and a terrible weed in the northwest United States—are much less pleasant. Leaves are alternate or opposite, have petioles and stipules, and many are hairy. Some have prominent dark patches, such as *Pelargonium zonale* and *Geranium phaeum*. The foliage can be simple or compound and, if divided, with either pinnate or palmate leaflet arrangement. Leaf margins are smooth or toothed. Shrubby species typically have prominently jointed stems and many of these are at least partially succulent.

Pelargonium zonale,
horseshoe geranium

Myrtaceae
The myrtle family

The trees and shrubs of the myrtle family are mostly evergreen, often have decorative peeling bark and exotic-looking flowers with many stamens. The Australian gum trees (*Eucalyptus*), tea trees (*Melaleuca*, *Leptospermum*), and bottlebrushes (*Callistemon*) are especially well known.

Size

With 5,500 species, this family includes major timber trees (*Corymbia*, *Eucalyptus*); fruits, such as guava (*Psidium*, *Acca*); and important spices, including cloves (*Syzygium aromaticum*) and allspice (*Pimenta dioica*). Several species (tea tree, eucalyptus) yield important essential oils.

Range

Primarily a family from tropical regions, garden-worthy plants hail from cooler southern South America (*Luma*, *Myrceugenia*, *Ugni*) and the Mediterranean (*Myrtus*). However, it's in Australia that the greatest diversity of hardy and decorative plants can be found.

Origins

Fossils of *Myrtaceae* fruit capsules from the Middle Eocene (around 45 million years ago) suggest that the family originated earlier, in the Late Cretaceous (90–70 million years ago).

Flowers

Myrtaceae flowers are radially symmetrical and can be solitary or clustered in various inflorescences. They usually have four or five sepals, four or five petals, and numerous stamens. In *Myrtus* and *Leptospermum*, the petals are prominent and attract pollinators. In *Callistemon* and *Melaleuca* they are small and green with attractive stamens. In *Eucalyptus* and some relatives, the petals (and sometimes the sepals) are fused together to form a cap (or calyptra) that falls off to reveal the brush of stamens.

Melaleuca cajuputi,
cajuput tea tree

*Flower in
section*

*Single
flower*

*Seed
capsules*

*Capsule in
section*

Callistemon speciosus,
Albany bottlebrush

Eucalyptus globulus,
blue gum
In eucalyptus flowers,
the petals and/or sepals are
fused together forming a
cap, which falls off to
reveal the numerous
white stamens.

Fruit

Fruits are fleshy berries (*Psidium, Myrtus*) or dry
capsules (*Eucalyptus, Callistemon*), some of which
open only after the application of fire.

Leaves

One of the most obvious characteristics of myrtle
family leaves is that they are often scented, with
spotlike glands dotted over the surfaces. This,
combined with their typically opposite
arrangement, makes them easy to identify.
The leaves are simple, occasionally alternate or
whorled, and have smooth margins. Petioles may
be long, short, or absent; in *Eucalyptus*, young
foliage is circular with no petioles, while mature
leaves are elongated with distinct petioles.

USES FOR THIS FAMILY

Gum trees make a bold statement in northern
landscapes, evoking the gray-green forests of the
bush and clashing with native local flora. Most
hardy gums have plain white flowers, but this is
more than made up for by their elegant form,
weeping foliage, and curious bark. *Eucalyptus*
grow rapidly, so choose with care; swamp gum
(*E. regnans*) is the second tallest tree in the world.
Where space is limited, control by pruning to
ground level (coppicing) or into a head
(pollarding), which also has the advantage of
stimulating the production of immature foliage.
Other myrtles worth growing include pineapple
guava (*Acca sellowiana*) and bottlebrushes, both
of which have striking stamens, while neat *Myrtus
communis* makes a lightly scented hedge. Chilean
myrtle (*Luma apiculata*) and guava (*Ugni
molinae*) are outstanding; the former has red
bark with contrasting dark green leaves, the
latter edible fruits.

Acca sellowiana,
pineapple guava

Onagraceae
THE EVENING PRIMROSE FAMILY

Onagraceae was named in 1836 for the genus *Onagara* (now known as *Oenothera*) by John Lindley, after whom the largest horticultural library in the world—the RHS Lindley Library in London, England—is named. The family contains a number of well-known garden plants, such as fuchsias, evening primroses, and clarkias.

Size

With 22 genera and 656 species, this is largely a family of herbaceous annual and perennial plants. Shrubby *Fuchsia* is a notable exception, and there are a few aquatic species, such as water purslane (*Ludwigia natans*).

Oenothera biennis,
common evening primrose

Range

Found virtually throughout the world, the *Onagraceae* is the most diverse in western North America. The majority of *Fuchsia* species originate from tropical and subtropical regions of Central and South America; the range of one hardy species (*Fuchsia magellanica*) extends to Cape Horn; it has also become an introduced weed in some parts of the world.

Origin

While the oldest fossils in the family belong to *Ludwigia*, it is the fuchsias that appear to be the closest living link to an ancestry that extends as far back as 93 million years.

Flowers

The four- or sometimes five-parted flowers are prominent features in almost every species, from the vivid yellow "cups" of evening primrose (*Oenothera*) to the delicate and starry, late summer flowers of whirling butterflies (*Gaura*). Anyone who has taken a close look at a fuchsia flower will notice that the ovary is inferior—i.e., in a position below the rest of the flower—swelling into the fruit upon pollination.

The petals, and sometimes also the sepals, are very brightly colored in order to attract pollinators, which are often birds but also bees and moths. Nearly all animals that pollinate *Onagraceae* flowers show specialized anatomy, such as the hummingbirds that are able to hover in front of each pendulous fuchsia flower while they insert their long beaks to reach the nectar.

Fruit

The fruit is sometimes a dry capsule or—in the case of *Fuchsia*—a small berry. These contain many tiny seeds that are either dispersed by animals (in the case of berries), or by the wind. The featherweight, fluffy seeds of fireweed or rosebay willowherb (*Chamaenerion angustifolium*) are so readily dispatched in the wind that the plant has become a weed in many parts of the world.

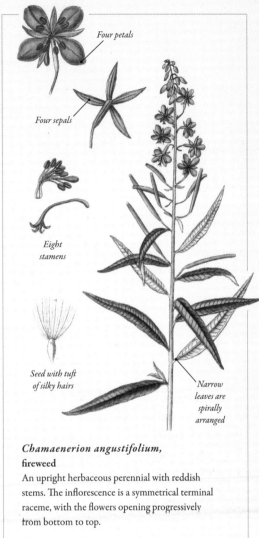

Four petals

Four sepals

Eight stamens

Seed with tuft of silky hairs

Narrow leaves are spirally arranged

Chamaenerion angustifolium,
fireweed
An upright herbaceous perennial with reddish stems. The inflorescence is a symmetrical terminal raceme, with the flowers opening progressively from bottom to top.

Fuchsia magellanica,
lady's eardrops

Leaves

Almost without exception, the leaves are simple in outline, and ovate to lanceolate in shape. In *Fuchsia* and *Epilobium* they are arranged in opposite pairs, but sometimes alternately arranged (*Chamaenerion*). Evening primroses often form basal rosettes, with leaves arranged spirally at ground level.

Sapindaceae
The maple family

Classical *Sapindaceae* was largely tropical and best known for exotic fruits, such as ackee, lychee, longan, and rambutan. Now much expanded, it also includes maples (*Acer*) and horse chestnuts (*Aesculus*), and ranges from woody trees, shrubs, and vines to herbaceous perennials.

Size

In its current configuration there are around 1,450 species in *Sapindaceae,* and many are useful.

Range

Sapindaceae are found almost worldwide, excluding the polar regions, and are also often absent from arid regions.

Origins

Fossil leaves related to modern *Sapindus* have been dated to 93 million years ago in the Cretaceous, though the modern genera did not appear until the Eocene.

Flowers

Sapindaceae blooms are extremely variable and can be ostentatious (*Aesculus*) or subtle (*Acer*). Most flowers are either male or female, sometimes with the genders on separate plants, though flowers containing both genders are not uncommon. There are four or five sepals, four or five petals (sometimes absent), and four to ten stamens, often with hairy stalks. The sepals can be fused into a tube, as with horse chestnut, or free like the maple, while the center of the flower has a distinctive nectar disc, often with the stamens attached to it.

Aesculus glabra,
Ohio buckeye

Individual flower

Koelreuteria paniculata,
golden rain tree

Fruit

There are many types of fruit in this family but perhaps the best known are the winged samaras of maple, the spiny capsules of horse chestnuts, and the inflated, papery capsules of rain trees (*Koelreuteria*).

Leaves

Almost all *Sapindaceae* have compound leaves with leaflets set in a pinnate arrangement (*Dipteronia, Sapindus*) or palmate fashion (*Aesculus*). In rare instances leaves are twice pinnate, reduced to three leaflets, or can be lobed (maples) or entire (*Dodonaea*). The leaflet edges can be toothed or smooth, and the petioles usually have somewhat swollen bases.

Acer palmatum,
Japanese maple

Rutaceae

THE CITRUS FAMILY

While it may seem strange that a family best known for its edible citrus fruits should be named for its most poisonous member—rue (*Ruta graveolens*)—this species is a good representative for the group because it shares many of its physical characteristics across the whole family.

Citrus aurantium,
Seville orange
The hesperidium (sectioned fruit) of the Seville orange is a hybrid between the pomelo and the mandarin. Its essential oils are used in perfume and the fruit itself is the key ingredient of marmalade.

Size

There are 150 genera and about 900 species of *Rutaceae,* consisting of shrubs, subshrubs, trees, and a few herbaceous perennials.

Range

Rutaceae are found all over the world in tropical and temperate regions. Although the largest share of species is found in the southern hemisphere, many of the best-known species seem to originate from Asia.

Origin

It is believed that the first citrus fruits evolved about 15 million years ago as small and edible berries. Related kumquats (*Fortunella*) and bitter orange (*Poncirus*) began to follow a separate path of evolution about 8 million years later.

Widespread in the fossil record since the beginning of the Paleogene (about 66 million years ago), *Rutaceae* fossils have also been found in Cretaceous sediments from several million years earlier. It is likely that the early evolution took place in North America.

Flowers

Regular, radially symmetrical, insect-pollinated flowers with four or five overlapping petals are the norm for the *Rutaceae*. Sometimes these have a strong fragrance. Colors range from white to yellow to green and deep red. Plants in the genus *Angostura* have irregular flowers.

Fruit

There are various fruit types within this family, from the capsules of rue (*Ruta*) to the drupes of *Skimmia*. Citrus fruit is actually a hesperidium (a modified berry with thick, leathery skin that contains aromatic oil glands), and its interior is split into sections, each packed with enlarged juice-filled cells and a number of seeds.

Ovary

Ovary in cross section

Immature seed capsule

Immature seed capsule in section

United carpels sharing single pistil

Ruta graveolens,
rue

This short, bushy perennial forms a mound of blue-green, ferny foliage, no more than 3 feet tall. Yellow flowers appear in early summer. The foliage is toxic and can cause blistering of the skin.

Individual flower, showing stamens

Stamen and seed

Seeds in section

Leaves

The genus *Ruta*, a good representative of the citrus family as a whole, has strong-smelling, compound pinnate or trifoliate leaves, and inflorescences of regular (radially symmetrical) flowers. The trees of *Citrus*, *Poncirus*, and *Phellodendron* are an obvious variation on this theme, some of which also have simple, rather than compound leaves. The aromatic leaves are the most distinctive feature of the *Rutaceae*; they are covered with oil glands that are small but visible as translucent black dots. The burning bush (*Dictamnus albus*) produces so much of its spice-scented volatile oil that on a very hot day a plant can easily catch fire.

USES FOR THIS FAMILY

The *Rutaceae* is home to all the citrus fruits, from lemons to oranges, tangerines to satsumas, and limes to grapefruit—a great addition to any subtropical or tropical garden. There are also some garden-worthy species for temperate climates, such as the Mexican orange blossom (*Choisya ternata*) and *Skimmia* × *confusa*.

Choisya ternata,
Mexican orange blossom

Malvaceae

THE MALLOW FAMILY

While perhaps best known for garden perennials, such as hollyhocks (*Alcea*), and shrubs including *Lavatera* and *Hibiscus*, the mallow family also extends to large tropical trees, such as kapok (*Ceiba*) and baobab (*Adansonia*), plus several important food (cacao, okra, cola) and fiber (cotton, jute) crops.

Size

In its traditional configuration, *Malvaceae* contained around 1,000 species, but recent DNA-based studies have led to its amalgamation with *Bombacaceae*, *Sterculiaceae*, and *Tiliaceae*. Thus it now comprises 5,000 species, including some well-known garden trees (*Tilia*), shrubs (*Abutilon*, *Fremontodendron*), and perennials (*Sidalcea*, *Althaea*, *Malva*).

Adansonia digitata,
baobab

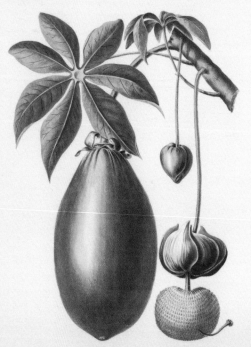

Althaea officinalis,
marsh mallow

Range

Distributed worldwide, though not in the polar regions, most of the diversity can be found in the tropics. Those species that occur in temperate regions are more likely to be shrubs or perennials, with a few exceptions, such as the lime/linden/basswood trees (*Tilia*).

Origins

Fossils dating back to the Late Cretaceous, around 68 million years ago, have been identified as coming from the erstwhile family *Bombacaceae*.

Flowers

Mallow blooms are often large, even gaudy, and make a striking addition to the yard. Each flower is radially symmetrical, has five sepals and five petals (rarely absent), and sometimes an additional whorl of sepal-like bracts (epicalyx) outside the sepals. While the petals are typically colorful, part of the attraction of *Malvaceae* flowers is the elaborate arrangement of the stamens—often numerous with their filaments fused together in a tube that surrounds the protruding style and stigma. While common, this arrangement is not universal; in *Tilia* the stamen filaments are only partially fused together, or not at all.

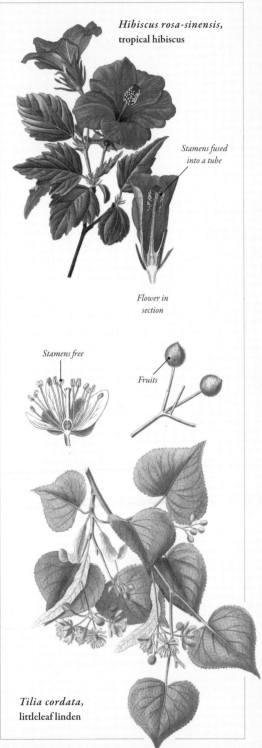

Hibiscus rosa-sinensis,
tropical hibiscus

Stamens fused into a tube

Flower in section

Stamens free

Fruits

Tilia cordata,
littleleaf linden

Lavatera phoenicea,
Canary shrub mallow

Stamen tube

Green epicalyx

Ovary in section

Gossypium barbadense,
Sea Island cotton

Seeds with silken hairs

USES FOR THIS FAMILY

Given that many species are restricted to the tropics, there are comparatively few for the temperate gardener, considering the great size of the family. In larger yards, linden trees make great specimens; *Tilia henryana* is especially pleasing with its distinctive toothed leaves and red new growth. Those aiming for a tropical look should search for a Chinese parasol tree (*Firmiana simplex*) or, where space is more limited, *Abutilon* × *suntense, A. megapotamicum*, or *Fremontodendron* make great exotic wall shrubs. In colder climates the tough-as-nails rose of Sharon shrub (*Hibiscus syriacus*) will bloom profusely, while cultivars of the cold-hardy rose mallow (*H. moscheutos*) will shock your friends with blooms the size of dinner plates.

Fruit

Several different fruits can be found in *Malvaceae*, and most are dry. Fleshy fruits occur rarely, as seen in the genus *Malvaviscus*. Most commonly, fruits are either capsules that split to release their seed, as in *Gossypium* (cotton) and *Hibiscus,* or schizocarps, which break apart into segments that each contain a few seeds, as seen in *Abutilon* and *Malva*. Others, like *Tilia*, have nutlike fruits that fall from the tree with an attached bract that acts as a parachute. Many *Malvaceae* fruits contain copious silken hairs, which aid in seed dispersal; those of cotton and kapok (*Ceiba*) are of commercial significance.

Hibiscus syriacus,
rose of Sharon

Ceiba pentandra,
silk-cotton tree
Silken hairs act as wings for windborne kapok seeds, when the large fruit capsule splits open.

Leaves

Perhaps the most obvious feature to look for in mallow leaves is the arrangement of veins, which is almost always palmate (arranged like the fingers of a hand). The leaves may be entire, toothed, lobed, or divided into leaflets (compound), but almost always with a palmate arrangement. When toothed, each leaf vein ends in a tooth.

The leaves are alternate along the stem and stipules are often present, though they're most easily seen on younger growth. Many mallows have hairs on the leaves and stems, and these are typically stellate; appearing like tiny stars when examined closely.

Cola acuminata,
cola nut tree
The bitter-tasting seeds of cola are used ceremonially in parts of West Africa. Rich in stimulants, including caffeine, they formed part of the original recipe for cola drinks.

Theobroma cacao,
cacao

Culinary Flavor

Several of the herbaceous species in *Malvaceae* are known as mallows and one, the marsh mallow (*Althaea officinalis*), has long been used as a food additive. Native to Europe, North Africa, and the Middle East, it was in the latter that mallow sap was first used to flavor a sweet snack called halva. The French developed the confection into its current form, though modern marshmallows lack the original flavoring. Other tasty *Malvaceae* include roselle (*Hibiscus sabdariffa*), whose flower gives flavor to hibiscus tea, and cola (*Cola acuminata* and other species), which provided the caffeine for the original Coca-Cola soda. Of course, the mallow family's biggest claim to fame is cacao (*Theobroma cacao*), a tropical tree whose colorful pods contain the seeds known as cocoa beans.

Cistaceae
THE ROCK ROSE FAMILY

There could be no better name than rock rose for this family that comprises beautifully flowering, modest shrubs suited to open rocky ground. Three of the genera—*Cistus*, *Halimium*, and *Helianthemum*—are common small-garden shrubs, valued for their short display of very showy flowers in early summer.

Size

This is a relatively small family of wiry-stemmed shrubs and subshrubs with pretty flowers, mostly suited to growing in dry sunny habitats. There are about 170 species split across just nine genera. The odd members of the family are *Tuberaria* and *Lechea,* which are not of a shrubby persuasion, being herbaceous perennials or annuals.

Helianthemum vulgare,
flower of the sun

Range

With the exception of three North American genera (*Crocanthemum, Lechea, Hudsonia*), the *Cistaceae* family is centered around the Mediterranean Basin, stretching into Asia and northern Europe. *Cistus, Helianthemum,* and *Halimium* are found in Mediterranean shrub communities, grassland, and open rocky sites, and although they are suited to the poor and dry alkaline soils of this region, they are fairly adaptable for garden cultivation, even in cold climates.

Halimium x *halimifolium,*
yellow-flowered rock rose

Origin

Like all of the *Rosid* group, the fossil evidence shows the family has ancestral links to the Late Cretaceous (about 70 million years ago), with later evolution of the *Cistaceae* through the Cenozoic Era (66 million years ago to the present day). Some of the genera and species within them may have only evolved fairly recently, which could explain why some of them readily hybridize within the genera (particularly *Cistus*), as well as between genera (× *Halimiocistus*).

Flowers

In the key genera of *Cistaceae*, the flowers are large, showy, and short-lived. They have a regular, radially symmetrical shape, with both male and female parts, and they are either solitary or borne in flower heads with cymes. *Lechea* are also known as pinweeds for their tiny flowers.

There are five sepals and five petals (*Lechea* has just three), which are overlapping and often crumpled in the bud. There are numerous stamens. The flowers bear a superficial resemblance to poppies. Most species have yellow flowers, sometimes with markings in the center. *Cistus* flowers are white or pink, while *Helianthemum* are white, yellow, pink, red, or orange.

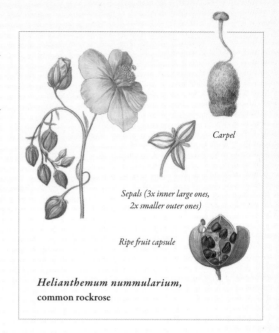

Carpel

Sepals (3x inner large ones, 2x smaller outer ones)

Ripe fruit capsule

Helianthemum nummularium,
common rockrose

Leaves

The leaves are opposite, fairly small, and simple in outline. *Cistus ladanifer* has strongly scented foliage.

Seeds

Any plant growing in the type of hot Mediterranean climate preferred by many *Cistaceae* has to be able to withstand the conditions. In the case of the rock rose family, it is the tiny, hard-coated seeds that are released in profusion when the fruit capsules dry in the sun. These seeds can sit in the soil for years and will only germinate after a fire has raged above ground, cracking the seed coating and allowing them to finally germinate once rain falls. Thus fire-scorched areas soon recover from the devastation.

Cistus × purpureus,
purple-flowered rock rose

Brassicaceae
THE CABBAGE FAMILY

Although there is great diversity in shape and size among plants in *Brassicaceae*, there is one almost constant feature that makes them all recognizable: their four-petalled, cross-shaped flowers for which the family was once named *Cruciferae*. The foliage often has a characteristic mustard scent.

Size

Altogether this is a large family with about 3,400 species across 321 genera, the majority of which are either annual or perennial herbaceous plants of various sizes. A few exceptions include subshrubs (*Alyssum, Iberis*) and a few tall shrubs, such as *Heliophila glauca* from South Africa.

Brassica rapa,
turnip

Range

Brassicas grow in most parts of the world but the greatest number of genera are found around the Mediterranean Basin and southwestern and central Asia. There are only a few representatives in the southern hemisphere and hardly any in the tropics. The most isolated genus must be *Pringlea,* which has just one species found on the Kerguelen Islands in the middle of the Southern Ocean.

Brassica cultivars are grown worldwide as vegetable crops. Unlike many other crops, there is an interesting amount of regional variation. Thus, Brussels sprout is traditionally seen very much as a British crop, cavolo nero an Italian one, and bok choy an Asian one.

Iberis umbellata,
candytuft
This plant is valued for its tight, umbellate flower clusters of small, four-petaled flowers. They are pollinated by bees and butterflies.

Origin

Brassicaceae began to flourish and diversify
66 million years ago after the great Cretaceous–
Paleogene mass extinction. *Brassica oleracea*,
the ancestral cabbage, was first cultivated about
8,000 years ago in coastal areas of western Europe.
Kale, cauliflower, Brussels sprouts, and kohl rabi
are all derived from this plant.

Flowers

Usually held in a raceme or corymb inflorescence,
the basic structure of the individual flowers is
remarkably constant across the family: four sepals,
four free petals in a cross-shape, and six stamens
(four long and two short). There are, of course, a
few exceptions, such as candytufts (*Iberis*), where
two of the petals are elongated.

Capsella bursa-pastoris,
shepherd's purse

Heart-shaped
seed pods

Seeds

USES FOR THIS FAMILY

The cabbage family contains many crop plants
that will be familiar to vegetable gardeners,
such as kale, flowering broccoli, and the roots
of rutabagas. Ornamental brassicas include
honesty (*Lunaria*), wallflowers (*Erysimum*),
night-scented stocks (*Matthiola*), and sweet
alyssum (*Lobularia*).

*Lunaria
annua,*
honesty

Fruit

All seed capsules in this family split open along
the line of the two valves. Besides this, there is
enormous variation in the shape of seed capsules,
and botanists rely on them as a way to tell apart
the different species in this family. When the
seed capsule is more than three times as long
as it is wide, it is called a siliqua; less than this,
it is termed a silicula. Examples range from the
round and flattened siliculas of honesty (*Lunaria
annua*), which dry into papery "moons," to
the heart-shaped silicula of shepherd's purse
(*Capsella bursa-pastoris*) and the long siliquae
of bittercress (*Cardamine*).

Amaranthaceae
THE AMARANTH FAMILY

A family of nonwoody, rapidly growing plants with tiny flowers massed into tight flower heads, which are often showy, as seen in the garden plants cockscomb (*Celosia argentea*) and love-lies-bleeding (*Amaranthus caudatus*). Red and yellow betalain pigments are a distinctive characteristic of this family, as seen in beets, the veins of chard, and the flowers of amaranths.

Size

The recent addition of *Chenopodiaceae* (goosefoot family), following a taxonomic review, makes this family quite large. There are 2,000 species, including beet, mangel wurzel, quinoa, and spinach, spread across 175 genera.

Range

Amaranthaceae is found all over the world. The main centers of diversity are in the African and American tropics, excluding plants formerly of *Chenopodiaceae*, which are centered around dry and warm-temperate areas (particularly in salty soils).

Origin

The fossil record of the whole *Caryophyllales* order (to which *Amaranthaceae* belongs) is poor. It is believed that this order emerged with the most primitive of the angiosperms, perhaps 100 million years ago during the Cretaceous. The *Amaranthaceae* is the order's largest family and will have almost certainly evolved much more recently.

At some point in the family's evolution, between 24 and 6 million years ago, there arose a modification in the photosynthetic pathway, which is seen in about 800 species of *Amaranthaceae*. Known as C4 plants, these are species that have a higher efficiency of water use during photosynthesis, enabling them to colonize drier habitats.

Salicornia europaea,
common glasswort

Celosia argentea,
plumed cockscomb

Amaranthus caudatus,
love-lies-bleeding
The name comes from from the tiny,
blood-red, petal-less flowers that bloom
in long tassels up to 20 inches long. Some
cultivated varieties come in different colors.

Flower head

Male flower

Female flower

Fruit and seed

Flowers

Typical of this family are conspicuous, spikelike
flower heads made up of small or tiny,
unremarkable bisexual flowers. This mass of
flowers can be flamboyant and ornamental. Often
the perianth is dry, membranous, and colorless,
and sometimes the flowers bear bracts, spines,
wings, or hairs.

Leaves

The leaves are mostly simple in outline and
alternate along the stem, but there is great variety
in shape and texture, and some species have
opposite leaves. In size, they range from the
minute scales of glasswort (*Salicornia europaea*)
to the big edible leaves of green amaranth
(*Amaranthus viridis*).

Sugar beet must be the most useful product of
Amaranthaceae. It comes from *Beta vulgaris*, a
species from which also derives cultivars such
as beets, mangel wurzels, and chard. Quinoa
and amaranth were important staples in
pre-Columbian South America.

Well-known ornamental species include
cockscombs, love-lies-bleeding, *Alternanthera*,
and *Iresine*. These are all very showy plants,
but their sensitivity to frost means that in
cool-temperate yards their use is restricted
to summer bedding or indoor display.

Beta vulgaris,
sugar beet

Stems

While the stems are usually nonwoody, there
are a few small shrubs in the family, such as spiny
hopsage (*Grayia spinosa*) from North America.
In a few species, the stems are quite fleshy, and
sometimes succulent, as seen in genus *Salicornia*.

Cactaceae

THE CACTUS FAMILY

This popular and well-known family of trees, shrubs, and herbs have succulent stems and often intimidating spines. Other spiny succulents in the spurge, milkweed, and even asparagus families are commonly called cacti, but their similarities result from their adaptation to comparable habitats.

Size

Cacti are extremely popular with hobbyists and virtually every one of the 1,210 species are in cultivation somewhere in the world.

Range

In the natural environment, cacti are almost entirely restricted to the Americas, stretching from Canada down to Patagonia. One species of mistletoe cactus (*Rhipsalis baccifera*) is also found in tropical Africa, Madagascar, and Sri Lanka, while prickly pears have become weeds in Australia and elsewhere.

Hatiora salicornioides, bottle cactus

Parodia ottonis, Indian head cactus

Origins

Soft-bodied succulents are probably the least likely of plants to form fossils, though a fossil has been found of what is believed to be a prickly pear (*Opuntia*) from the Eocene (53–48 million years ago). Given their restricted range, it seems likely they evolved in South America after it split from Africa during the Cretaceous.

Flowers

The flowers of the cactus family are usually solitary, emerge from the areoles (see Leaves section) and can have a covering of spines or hairs. The blooms are radially symmetrical (bilateral in Christmas cactus) with abundant, largely similar sepals and petals. Cactus flowers come in many lurid colors, though night-flowering species are usually white. Stamens are numerous.

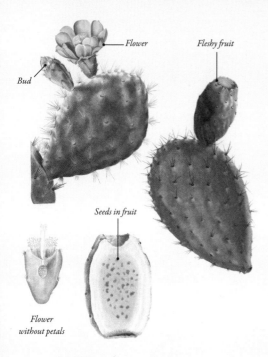

Flower

Fleshy fruit

Bud

Seeds in fruit

Flower without petals

Nopalea cochenillifera,
cochineal cactus
This prickly pear-type cactus is one host of the parasitic cochineal insect, from which the carmine-colored dye of the same name is extracted.

Fruit

The fruits are fleshy berries, rarely dry, filled with numerous small black seeds. Some fruits have spines, hairs, or bracts on the skin.

Leaves

Most cacti have no leaves, but there are exceptions. *Pereskia* are woody shrubs with simple, alternate leaves; they are thought to be the most primitive cacti. Some prickly pears produce small leaves on new shoots, but these quickly wither. It's the succulent stems that perform the photosynthetic duties of leaves, and these may be cylindrical, globe-shaped, flattened, or pad-like, often with ridges allowing them to swell with water. On the

stem surface, often along these ridges, are squat, hair-covered stems called areoles. These produce spines (which are modified leaves), flowers, and the irritating hairs (known as glochids) found on prickly pears. Cactus spines can be thick, ridged, hooked, or flimsy, while some species are completely unarmed.

Cylindropuntia imbricata,
cane cholla

USES FOR THIS FAMILY

Most gardeners would instinctively consider cacti to be unsuitable for cultivation in temperate gardens, but several desert species are highly tolerant to cold temperatures, as long as they don't get too wet in winter. Many prickly pears (*Opuntia*) and chollas (*Cylindropuntia*) are extremely hardy, forming vicious-looking shrubs for sunny borders, while diminutive *Maihuenia poeppigii* is perfect for a rock garden or trough. During summer, hang Christmas and Easter cacti (*Schlumbergera*, *Hatiora*) in garden trees to encourage abundant blooms when brought indoors in the fall.

Pereskia bleo,
rose cactus

Caryophyllaceae
THE CARNATION FAMILY

Carnations and pinks are known the world over for their white, pink, or red flowers with frilly or deeply lobed petal tips. While a few species are shrubby, such as *Acanthophyllum*, and a few have succulent stems, such as *Honckenya*, the family is best known for its herbaceous members.

Size

Caryophyllaceae is a relatively large family of 85 genera and 2,630 species. These are mainly nonwoody annuals or perennials that are short in stature and die back to ground level during winter. The family is divided into three subfamilies, based on their flower anatomy. The first two are well known, easily distinguished from each other and

Acanthophyllum
spinosum

Honckenya peploides,
sea purslane

easily recognized: subfamily *Alsinoideae* contains species where the sepals are free from each other (*Arenaria, Stellaria, Cerastium, Sagina*); subfamily *Silenoideae* contains species where the sepals are joined together, often forming a tube or bladder (*Silene, Dianthus, Gypsophila, Agrostemma, Lychnis*). Many members of this second subfamily cross-breed readily, a trait that has been massively exploited in genus *Dianthus* to create thousands of cultivated varieties with showy flowers.

Although the third subfamily, *Paronychioideae*, does not contain any genera of horticultural note, it still contains notable genera such as spurreys and sea spurreys (*Spergula, Spergularia*) and manyseeds (*Polycarpon*). It is also more varied than the first two subfamilies.

Agrostemma githago,
corn cockle

The most economically important member of
the *Caryophyllaceae* is the carnation (*Dianthus
caryophyllus*), cultivated worldwide as both a cut
flower and as a border perennial for gardens.
Many hundreds of novel cultivars are produced
across the world on an industrial scale each year.
Other species of *Dianthus* are widely cultivated
in gardens, usually sold as "pinks," as well as
campions, baby's breath (*Gypsophila*), corn cockle
(*Agrostemma githago*), fire pinks (*Silene virginica*),
and snow-in-summer (*Cerastium tomentosum*).

Dianthus caryophyllus,
border carnation

Range

Members of *Caryophyllaceae* can be found in
all parts of the temperate world, even on tropical
mountain tops. A few weedy species, such as
common chickweed (*Stellaria media*) and
mouse-ear chickweed (*Cerastium*), grow wild and
have also become almost universal in gardens.
Corn cockle (*Agrostemma githago*) was once a
common weed of grain fields, but modern farming
practices have all but removed it from the
agricultural landscape.

The center of distribution is the Mediterranean
Sea and neighboring parts of Europe and Asia,
where all the larger genera can be found, such
as campions (*Silene*), pinks and carnations
(*Dianthus*), and sandworts (*Arenaria*). There are
about 20 genera in North America and just a few
in the southern hemisphere. One species of note is
Antarctic pearlwort (*Colobanthus quitensis*),
which is one of only two flowering plants that call
the continent of Antarctica their home—the other
is Antarctic hairgrass (*Deschampsia antarctica*)
from the family *Poaceae* (see pages 96–99).

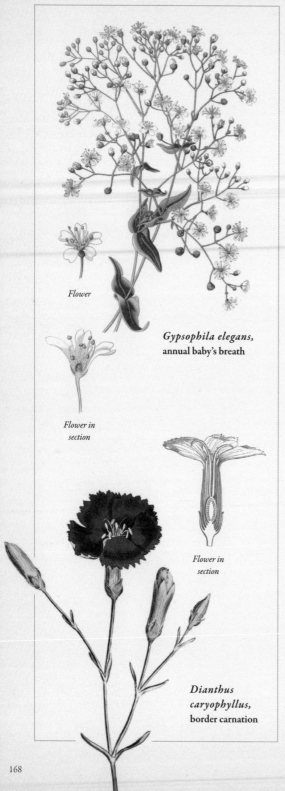

Flower

Gypsophila elegans,
annual baby's breath

Flower in section

Flower in section

Dianthus caryophyllus,
border carnation

Carpel, stamens, and petal

Lychnis flos-cuculi,
ragged robin

Origin

Despite the carnation family's success in modern botanical history, there is very little trace of its evolution available in fossil records. Botanists believe that it is very closely related to the amaranth family (*Amaranthaceae*—see pages 162–163) and therefore shares a common ancestry. The two families may have diverged approximately 50–40 million years ago, during the Eocene.

Flowers

Caryophyllaceae flowers are so recognizable it is often not necessary to examine their anatomy to confirm identification. They either bloom singly or in branched, cymose flower heads. Some of these carry just a few flowers, as seen in *Lychnis*; others are prolific, such as *Gypsophila* (baby's breath).

The flowers are radially symmetrical with four to five petals and the same number of sepals. The sepals

There are usually ten stamens appearing in one or two whorls, or sometimes just an equal number or fewer, as in *Stellaria media*. On pollination, the fruit matures as a dry capsule bearing numerous seeds, opening through valves at the top.

Leaves

The leaves are always simple and entire, and nearly always opposite each other on the stem. The stem nodes, where the leaves are joined, are swollen, and the bottoms of each leaf pair often join around them to form a united (or perfoliate) base.

The small leaves of *Cerastium* are often textured with many tiny, feltlike hairs, which is what gives this plant its common name, "mouse-ear."

Lychnis chalcedonica,
Maltese cross
Named for its cross-like flowers, which are made up of five lobed petals, 10 stamens, and a superior ovary. They are borne in clusters of up to 50 flowers.

Stellaria media,
chickweed

Silene vulgaris,
bladder campion

are often prominent and may all be free or united together, sometimes forming a tube (*Dianthus*), or inflated, as seen in bladder campion (*Silene vulgaris*).

There are a range of flower colors: white (*Gypsophila paniculata*), pink (*Dianthus caryophyllus*), red (*Lychnis chalcedonica*), and almost purple-pink (*L. coronaria*). There is no blue pigmentation in this family. Yellows are sometimes seen, for example in *Dianthus knappii*. The many varied colors are often mixed by plant breeders to create stunning flowers for florists.

The apices (tips) of the petals are often fringed, notched, or cut, or have deep or shallow lobes. In a few cases, the lobes are so severe that it looks as though there are twice as many petals as there are (*Stellaria media*). This effect can be very eye-catching, for example in ragged robin (*Lychnis flos-cuculi*). In a few rare cases, the petals are absent.

Droseraceae
THE SUNDEW FAMILY

For many gardeners, an obsession with plants began in childhood with the gift of a Venus flytrap (*Dionaea muscipula*). However, interest in this macabre family of carnivores, which also includes sundews (*Drosera*) and waterwheel (*Aldrovanda*), is not restricted to children.

Size

Almost all 105 species in this family are sundews, characterized by sticky leaves that bend over when a bug is caught. The two exceptions are Venus flytrap and the aquatic waterwheel plant (*Aldrovanda vesiculosa*), both of which have traps that snap shut on unsuspecting prey.

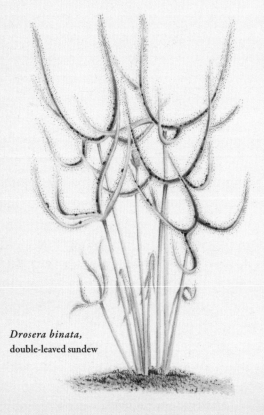

Drosera binata,
double-leaved sundew

USES FOR THIS FAMILY

Carnivores make great patio plants and will help to control mosquitoes and other biting insects. Often confined to windowsills, most sundews and Venus' fly traps are best grown outside, where they'll receive the high light levels they crave. Grow these bog plants in a container without a hole in the bottom in a peat-free low-nutrient specialist compost. Never apply fertiliser. Keep the soil wet, though ensure the water level remains below the crowns. Move the pot into a shed or garage in winter.

Range

Sundews are found on every continent except Antarctica, though they are especially diverse in Australasia. They typically occur in bogs and wetlands, as does *Dionaea*, which is restricted to North and South Carolina in the United States. The aquatic waterwheel ranges from Europe to Australia.

Origins

Fossilized pollen dating back 48–34 million years to the Eocene has been identified as coming from *Droseraceae*.

Drosera stenopetala,
New Zealand sundew

Dionaea muscipula,
Venus flytrap

Aldrovanda vesiculosa,
waterwheel plant

Drosera rotundifolia,
round-leaved sundew

Flowers

Flower clusters are held above the plant on long
stalks to preserve pollinating insects from the traps
below. Each bloom has five sepals, usually fused at
the base, and five free petals in shades of pink, red,
or white. Five stamens are typical, though *Drosera*
species can have 20 or more. In the case of
waterwheel, flowers are solitary.

Leaves

Sundew leaves, with their sticky tentacles, are
usually arranged in rosettes, either flat upon
the ground or upright. They can be simple or
branched, with new leaves unrolling like fern
fronds. Most species are herbs, some are tuberous,
and a few make upright shrubs, such as *Drosera
magnifica,* which was discovered in Brazil only in
2012. The foliage of Venus flytrap is spoon-shaped
with fringed heads, and red-colored inner faces
with trigger hairs that spring the trap. *Aldrovandra*
is free-floating, without roots, and has whorls of
filamentous leaves tipped with small traps.

Predatory Plants

Carnivory in plants probably evolved so species
growing in poor soils could supplement their
nutrient requirements. Nitrogen is especially
sought after and the protein-rich bodies of animals
are an abundant source. Venus flytraps attract
insects with a sugary lure and snap shut only if the
surface hairs are stimulated twice. Once shut, the
trap is sealed and digestive enzymes begin to break
down the insect's body.

Polygonaceae
THE RHUBARB FAMILY

This is an easy family to identify once you know what to look for. Key traits of *Polygonaceae* include swollen nodes along the stems (*Polygonum* translates as "many knees"), small flowers in spikelike heads, and triangular seeds. Gardeners will recognize common family members, such as rhubarb, sorrel, and the infamous Japanese knotweed.

Size

There are approximately 1,200 species across 46 genera. Most are nonwoody perennials; there are also some trees, shrubs, and climbers. The family does not include giant rhubarb (*Gunnera manicata*), which belongs to its own family: *Gunneraceae*.

Polygonum was once the largest genus, but now many species have been reorganized into other genera, for example Russian vine (*Fallopia baldschuanica*), bistort (*Persicaria*), and buckwheat (*Fagopyrum*). This genus is often confused with Solomon's seal (*Polygonatum*), which belongs to an entirely different family: *Asparagaceae* (see pages 88–89).

Range

Polygonaceae is predominantly a family of the northern hemisphere, and its constituent genera can be broadly grouped into three climatic groups: tropical and subtropical, desert and semidesert, and temperate regions. Mexican coral vine (*Antigonon leptopus*) and Australian maidenhair vine (*Muehlenbeckia complexa*) are two well-known examples of the first group, and California buckwheat (*Eriogonum fasciculatum*) an example of the second. From temperate regions come rhubarbs (*Rheum*), sorrels (*Rumex*), and buckwheats.

Origin

Based on fossil pollen, *Polygonaceae* are estimated to be about 60 million years old. The unusual distribution of *Muehlenbeckia* around the southern hemisphere suggests the family evolved before the separation of the supercontinent Gondwana.

Flowers

Typical of this family are spikelike flower heads made up of lots of small greenish, white, or pinkish flowers. Individually, the flowers are radially symmetrical and usually have both male and female parts. There are no true petals, but instead three to six sepals; in a few species, these are relatively large and colorful.

Fallopia baldschuanica, silverlace vine

Fruit

After pollination, the sepals may increase in size and become membraneous around the fruit. They are often persisting and sometimes even more eye-catching than the flowers. The fruit itself is a distinctive brown or black triangular nut, and the angles may be winged.

Leaves

The leaves are alternately arranged and simple in outline, often with heart-shaped bases. They range from the large, rounded, and textured leaves of rhubarb to the narrower lanceolate to oval leaves of *Rumex* and *Persicaria*. Characteristic of the family is a membraneous sheath around the joint of the leaf stalk and the stem, known as an ocrea.

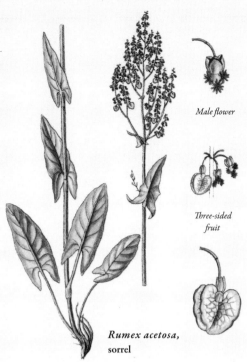

Male flower

Three-sided fruit

Rumex acetosa, sorrel

Fallopia japonica, Japanese knotweed

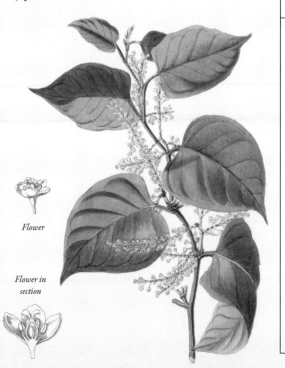

Flower

Flower in section

USES FOR THIS FAMILY

There are a number of cultivated species in the rhubarb family, from the moisture-loving *Rheum palmatum* (Chinese rhubarb), to the popular border perennial *Persicaria amplexicaulis* (red bistort) and the vigorous climber *Fallopia baldschuanica*. Edible members include sorrel (*Rumex acetosa*) and rhubarb (*Rheum* x *hybridum*).

Rheum x **hybridum,** rhubarb

Cornaceae
THE DOGWOOD FAMILY

This family of trees and shrubs includes several genera of horticultural importance, notably the dogwoods (*Cornus*), dove tree (*Davidia*), and tupelos (*Nyssa*). Some scientists, however, separate the latter two into a separate family, *Nyssaceae*.

Size

Of the 80 species in *Cornaceae*, about half belong to the dogwood genus *Cornus*. Also known as cornels, they are extremely variable, ranging from large trees (*Cornus controversa*) to ground cover (*C. canadensis*). Some, such as *C. florida*, have decorative floral bracts, while others, such as such as *C. kousa*, are known for their attractive fruits. *C. sericea* and *C. sanguinea* are known for colorful winter stems.

Alangium chinense,
begonia-leaf alangium

Range

Cornaceae are widely distributed, with some species found on every continent but Antarctica.

Origins

There is fossil evidence of *Cornaceae* dating back to the Cretaceous (84–72 million years ago). The abundance of fossilized seeds suggests this family was historically widespread.

Flowers

The flowers, which can be male, female, or both, are clustered in inflorescences at the tips of stems. White leafy bracts enclose the inflorescences of *Davidia* and some *Cornus,* whose bracts can also be pink or red. Sepals are often toothlike or absent while petals are small and free. There are usually four sepals and petals in *Cornaceae*, and five in *Nyssaceae*.

Nyssa aquatica,
water tupelo

Fruit

Fruits are fleshy, sometimes colorful, with a hard-coated seed inside. In *Cornus kousa,* and a few other dogwoods, several fruits fuse together to form an alien-looking, warty compound fruit. The dove tree produces hard, green nuts that turn purple when ripe.

Leaves

Foliage can be without stipules, evergreen, or deciduous, opposite or, less often, alternate. The tupelos and some dogwoods have pleasing foliage color in the fall, while several *Cornus* have been bred for variegated leaves. These are usually simple and entire, though *Davidia* leaves are toothed and those of some *Alangium* are lobed. Dogwood leaves are readily identified by tearing them in half; the two halves remain connected at the veins by silky threads.

USES FOR THIS FAMILY

With its dangling handkerchief-like inflorescences, a mature *Davidia involucrata* is an arresting sight, though it's only suitable for larger yards. Likewise, where space allows, the scarlet fall foliage of *Nyssa sinensis* and the layered branching of *Cornus controversa* are attractive. Smaller evergreen dogwoods, such as *C. kousa* and *C. angustata*, are more delicate; their profuse summer blooms contrast with dark, glossy foliage, and later develop into colorful fruits. Alternatively, winter-flowering Cornelian cherry (*C. mas*) blooms before the deciduous leaves emerge, and the red, cherry-like fruits are tasty. And if even a small tree is too big, plant a palette of dogwoods with colorful winter stems (available in red, orange, yellow, acid green, and black); they make the perfect foil for spring bulbs.

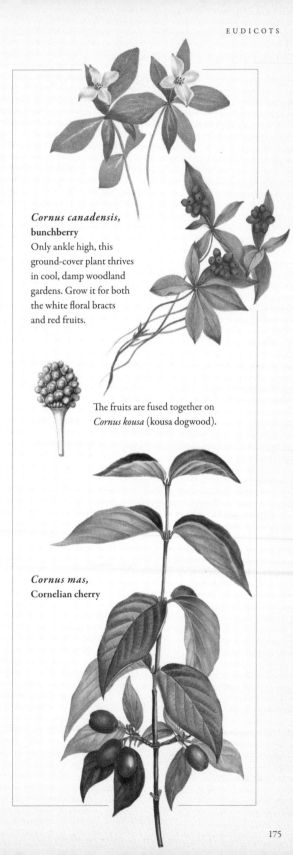

Cornus canadensis, **bunchberry**
Only ankle high, this ground-cover plant thrives in cool, damp woodland gardens. Grow it for both the white floral bracts and red fruits.

The fruits are fused together on *Cornus kousa* (kousa dogwood).

Cornus mas, Cornelian cherry

Hydrangeaceae
THE HYDRANGEA FAMILY

Though best known for the hydrangeas, this family also includes several other popular shrubs (*Philadelphus*, *Deutzia*), climbers (*Schizophragma*, *Pileostegia*), and herbaceous perennials (*Kirengeshoma*, *Deinanthe*).

Size

This small family of around 220 species was long considered part of *Saxifragaceae*, but examination of their DNA has revealed they're actually closer to the dogwoods in *Cornaceae*.

Range

Showing the greatest diversity in eastern Asia and North and Central America, a handful of species of *Hydrangeaceae* reach into South America, Europe, and the Pacific islands.

Origins

Fossilized flowers from the Late Cretaceous (around 90 million years ago) show a strong resemblance to *Hydrangeaceae*, though they may also represent *Saxifragaceae*. Modern genera are widespread from the Oligocene (34–23 million years ago) onwards.

Flowers

This family's flowers are usually clustered into inflorescences, and most blooms have both male and female parts. There are four or five sepals, usually fused together, and four or five petals, again often fused. Stamens range from two to numerous. In many hydrangeas, some of the flowers are sterile with enlarged, often colorful, petal-like sepals. In mophead hydrangeas the inflorescences are composed almost entirely of sterile blooms, while in lacecaps sterile flowers surround the central fertile blooms.

Philadelphus "Belle étoile," mock orange

Dichroa febrifuga, Chinese quinine

Sterile flower

Fertile flower

Flowers
are blue in
acidic soils

Flowers
are pink in
alkaline soils

Hydrangea macrophylla,
bigleaf hydrangea

An unusual characteristic of bigleaf hydrangea (*Hydrangea macrophylla*) is its ability to change flower color depending on soil pH. In acidic soils aluminum ions become soluble and, once absorbed, change the floral pigments to blue or purple. In alkaline soils the ions are insoluble and unavailable to the plants, whose flowers are then pink or red. Thus, hydrangea is a living litmus test for soil pH. Not all cultivars can change color; white blooms, for example, will always be white. Adding aluminum sulfate to an alkaline soil will encourage blue blooms, while adding lime to acidic soil encourages pink flowers. However, if your soil is very strongly acidic or alkaline then it can be difficult to change, and it is better to grow your hydrangeas in pots of suitable soil.

USES FOR THIS FAMILY

Though hydrangeas have many uses around the garden, other members of the family are worthy of cultivation. For its large white flowers with numerous golden stamens, grow evergreen *Carpenteria californica* in sheltered courtyards and gravel yards. The waxy blooms of *Kirengeshoma* and *Deinanthe* dazzle in a woodland garden, while evergreen vine *Pileostegia viburnoides* will happily clothe a shady wall or dead tree.

**Kirengeshoma
palmata,**
palmate kirengeshoma

Leaves

In most species the leaves are deciduous, opposite, and with toothed or smooth margins. *Hydrangea* itself includes several exceptions, such as the lobed leaves of *H. quercifolia* and the evergreen foliage of *H. integrifolia* and *H. seemannii*. Other evergreens include *Carpenteria*, *Dichroa*, and some *Philadelphus*, while the leaves of *Kirengeshoma* are lobed.

Ericaceae
THE HEATHER FAMILY

Hugely useful in the garden, this family of mainly woody plants includes the heathers (*Calluna, Erica, Daboecia*), azaleas, rhododendrons, wintergreens (*Gaultheria*), *Pieris*, and mountain laurels (*Kalmia*). Commercially significant crops include blueberries and cranberries (*Vaccinium*).

Size

One of the larger families, the *Ericaceae* contains over 3,850 species. Within this great diversity are many small genera with one or two species, and three titans: *Rhododendron* (with 1,000 species), *Erica* (850 species), and *Vaccinium* (500 species). It should be noted that azaleas are included within *Rhododendron*.

Range

Ericaceae is absent from Antarctica and from most tropical lowland forest. It is common on tropical mountains, in southern Africa, and in the cooler parts of eastern North America and eastern Asia.

Origins

Earliest evidence of this family dates to the Late Cretaceous (about 90 million years ago). Fossils suggest that *Ericaceae* was once more diverse in Europe, which was home to genera now restricted to Asia and/or America.

Flowers

Great floral diversity is encompassed by this family, so keep in mind that there are always exceptions to the basic forms described here. Flowers usually include both male and female parts, are clustered in inflorescences, and are somewhat pendulous. In *Rhododendron*, numerous (often sticky) bracts protect the unopened flowers. There are four or five

Flower in section

Rhododendron canescens,
mountain azalea

Calluna vulgaris,
ling heather
The only member of the genus *Calluna*, ling heather is found across Europe and western Asia, forming the dominant species on moor and heathlands.

sepals, free or fused at the base, and four or five petals, sometimes free, though usually fused together to form a tube, bell, or urn. The stamens are in whorls of four or five and the pollen is released from the anthers by way of pores at the tips.

Fruit

Fruits are typically dry capsules, though fleshy fruits, such as blueberries, are not uncommon.

Leaves

Most *Ericaceae* have evergreen, alternate leaves and no stipules. Some species, such as many azaleas, are deciduous, while opposite and whorled leaf arrangements are also known. Leaf margins are entire, toothed, or curled under, and some species (including many rhododendrons) have dense hair or scales on the lower surfaces.

Vaccinium corymbosum,
highbush blueberry

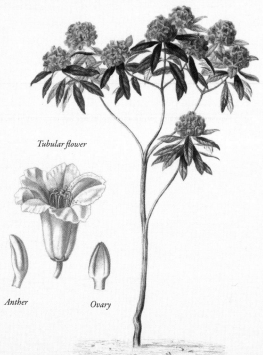

Tubular flower

Anther   *Ovary*

USES FOR THIS FAMILY

This family is often found in areas with nutrient-poor, acidic soils, such as heathlands (named for heather) and peat bogs. They survive by binding their roots to fungal threads in the soil, creating a symbiotic relationship known as mycorrhiza. The fungus greatly improves the roots' ability to extract soil nutrients and, in return, the plant gives to the fungus sugars manufactured by photosynthesis. Most *Ericaceae* require acidic soils, so if your soil is alkaline they are unlikely to thrive. To get around this problem you can grow small species in containers using an acidic compost named for the family (ericaceous). If you're not sure what kind of soil you have, buy a pH test kit or observe neighboring yards to see what thrives.

Rhododendron arboreum,
treelike rhododendron
The genus *Rhododendron* is the largest in *Ericaceae* and includes large trees like this, through to tiny creeping alpine species, though most are shrubs.

Primulaceae

THE PRIMROSE FAMILY

This family of mostly small- or medium-size, nonwoody herbaceous perennials includes a few annuals, such as scarlet pimpernel (*Anagallis arvensis*). There is a general liking among plants in this family for moist, damp, or boggy soils, and a few species—such as water pimpernel (*Samolus*) and featherfoil (*Hottonia*)—are aquatic. Species from the former *Myrsinaceae* family are woody trees and shrubs.

Size

What was once a modest family of 28 genera has, with the inclusion of former families *Myrsinaceae* and *Theophrastaceae*, grown into a larger unit of 60 genera and 2,575 species. Thus, *Primulaceae* now includes coralberry (*Ardisia*), sowbread (*Cyclamen*), yellow loosestrife (*Lysimachia*), and colicwood (*Myrsine*).

Cyclamen hederifolium, ivy-leaved cyclamen

Range

This family has a broad range, mostly circling the temperate northern hemisphere. Outliers in Africa, South America, New Zealand, and the tropical zones tend to be newcomers from former families *Myrsinaceae* and *Theophrastaceae*.

Primula auricula, primula

Cyclamen is native to Europe and North Africa, extending into western Asia, and nearly half of all *Primula* species are native to the Himalayas. Shooting stars (*Dodecatheon*) are from North America.

Origin

The *Ericales* order, to which *Primulaceae* belongs, is well represented in fossil records from the Late Cretaceous (approximately 90–80 million years ago), but fossil evidence is too sparse to provide much information about this family's origin.

Flowers

The flowers are typically five-parted, with five petals, five sepals, and five stamens. Note that *Lysimachia* is an exception, having six sepals. The sepals and petals are united to form two separate tubes. The stamens are joined to the corolla, aligned opposite each petal. In the genera *Samolus* and *Soldanella* there are also five staminodes (stamens without anthers) that alternate with the petals.

Except for the Mediterranean genus *Coris*, the flowers in this family are regular in shape (radially symmetrical). In some species, such as *Cyclamen* and *Dodecatheon*, the petals are reflexed backward.

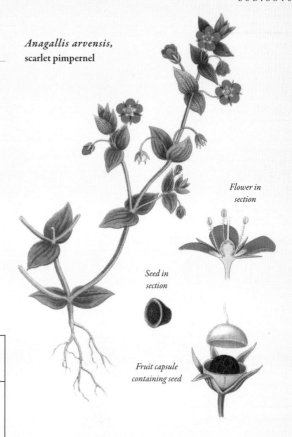

Anagallis arvensis,
scarlet pimpernel

Flower in section

Seed in section

Fruit capsule containing seed

The flowers are either carried singly or in umbellate, racemose, or paniculate inflorescences on long, leafless stalks.

In genus *Primula*, there are two distinct arrangements of the style and the stamens, known as pin and thrum; pin-eyed flowers have a style longer than the stamens, looking like there is a pin in the mouth of the flower tube, while thrum-eyed flowers have stamens longer than the style.

Leaves

With the exception of genus *Hottonia*, with its pinnate aquatic leaves, the leaves in *Primulaceae* are simple and undivided. In many species, particularly *Primula*, the leaves form a basal rosette at ground level. Glandular hairs are often present on the stems as well as the leaves.

USES FOR THIS FAMILY

Primulaceae contains many ornamental plants, valued for their pretty flowers. The list of garden-worthy species includes most of the primulas and cyclamen species, as well as shooting stars, creeping Jenny (*Lysimachia nummularia*), and snowbells (*Soldanella*).

Dodecatheon meadia,
shooting stars

Theaceae

THE CAMELLIA FAMILY

As referenced in the family name, *Theaceae* is best known as the source of tea leaves, from *Camellia sinensis*. A family of trees and shrubs, many are important ornamentals, including *Stewartia*, *Franklinia*, *Pyrenaria*, *Polyspora*, and the numerous species and cultivars of *Camellia*.

Size

Theaceae now includes around 240 species, though recently 330 species were removed into the new family *Pentaphylacaceae*, which differs in its fleshy fruits and distinct petals and sepals.

Range

This family can be found in tropical and temperate areas of eastern and southeastern Asia, and eastern North America through the Caribbean to South America. It is absent entirely from Europe, Africa, Australia, and Antarctica.

Stewartia ovata var. grandiflora,
stewartia

Origins

Most *Theaceae* fossils are relatively recent. Those from the Late Cretaceous (around 70 million years ago) await conclusive identification; they resemble the genus *Schima*, though they also resemble *Pentaphylacaceae*.

Flowers

Camellias and their relatives have solitary blooms, which appear from the leaf axils. Each is made up of several bracts, then five sepals and five petals, all of which can intergrade and appear similar. Many camellia cultivars have double flowers with numerous petals and few or no stamens; typical blooms have abundant golden stamens. Most *Theaceae* have white flowers, though camellia blooms range from white to pink to red and, in rare instances, yellow (*Camellia petelotii*).

Camellia japonica,
camellia
Native to China and Korea as well as Japan, this species is the parent of many popular garden hybrids and cultivars.

Camellia sinensis,
tea plant

*Ovary in
section*

*Flower in
section*

*Fruit from
above*

*Fruit from
below*

*Seed and
cotyledons*

Leaves

Most *Theaceae* are evergreen, except for
Franklinia and most *Stewartia*, which are
deciduous species and often produce fine fall
coloration. Their leaves are alternate, simple,
and toothed, with a deciduous gland at the tip
of each tooth. Several species of *Stewartia* have
attractive mottled bark that comes into its own
once the leaves have fallen in winter.

USES FOR THIS FAMILY

Like their relations in *Ericaceae*, camellias and
their kin prefer soils that are acidic and preferably
rich in organic matter. They perform well in a
woodland setting, while smaller species are suited
to container culture. For something different,
plant a hedge using camellias; though not as
fast-growing as some hedging, they'll form a
thick evergreen screen with the benefit of blooms.
Train fall-flowering *Camellia sasanqua* against a
drab north- or east-facing wall, or site *Franklinia*
as a border focal point with large white blooms
and excellent fall leaf color.

Camellia sasanqua,
fall-flowering camellia

Fruit

Fruits are dry capsules (rarely fleshy) that open
to release winged or wingless seeds. In the garden
camellias rarely fruit, and one reason is that many
cultivars are sterile. If you do have fertile plants,
which usually have single flowers with lots of
stamens, and some hardworking bees, then the red,
apple-like fruits may form. Leave them on the
plant until they split open, then collect seed to
grow more plants. The seeds are rich in oil and
those of *Camellia oleifera* are harvested to extract
tea oil, for use in cooking and cosmetics.

Convolvulaceae

THE MORNING GLORY FAMILY

Also known as the morning glory family, *Convolvulaceae* is made up largely of climbing, twining, and trailing plants, some herbaceous and some woody. Most gardeners will recognize morning glory (*Ipomoea tricolor*), or have dealt with the dreaded bindweed, and therefore will be familiar with this family.

Size

There are 52 genera and 1,650 species. The largest families are *Ipomoea* (morning glory), with more than 500 species, and *Convolvulus,* with about 230 species. Dodder (*Cuscuta*) is an unusual genus of about 150 species of parasitic plants. There are also a few shrubs and trees. *Evolvulus*, for example, is a genus of about 100 species of annuals, perennials, and shrubs with nontwining stems.

Range

The family is found in all parts of the world, across temperate and tropical regions, in a diverse number of habitats, from richly vegetated, to dry and sparse semidesert regions, including sand dunes, where you will find *Ipomoea pes-caprae*. *Ipomoea sagittata* (commonly known as saltmarsh morning glory) grows on the edges of salt marshes and water spinach (*I. aquatica*) grows in fresh water.

Origin

Convolvulaceae has only a scattered fossil record, so it is difficult to trace its exact origin. It belongs to the order *Solanales*, which botanists speculate evolved during the Cenozoic Era (66 million years ago).

Ipomoea tricolor,
morning glory
Grown for its trumpet-shaped flowers, the stems twine and climb over supports and other plants. The leaves are spirally arranged.

Calystegia sepium,
hedge bindweed

*Flower in
section*

*Ripe fruit
capsule*

Seed in section

Leaves

The family's name derives from the Latin
convolvere, which means "to wind" or "to wrap,"
in reference to the long and twining stems. These
stems sometimes exude a milky sap when cut.
The simple leaves are arranged alternately.

In most species the leaves and stems are green,
sometimes tinted pink or purple. *Cuscuta* are a
parasite on other plants, and as a result the stems
are often yellow, orange, or red in color,
with minute scales in place of leaves.

*Evolvulus
arbuscula,*
shrubby evolvulus

Flowers

The showy flowers in this family are radially
symmetrical, with the five petals united into a
funnel or trumpet shape. The slightly twisted
corolla is typically blue, pink, or white—
sometimes yellow, cream, red, or scarlet—and
often has a star pattern radiating from the center.
There are usually five stamens and a single pistil
with a superior ovary.

The sepals are usually distinct and, unlike
the petals, not united. Occasionally, the flower
bud is enclosed by bracts, which persist around
the base of the calyx during and after flowering,
as seen in hedge bindweed (*Calystegia sepium*).
The flowers themselves are either carried singly
or in cymose flower heads.

Solanaceae

THE NIGHTSHADE FAMILY

With its very distinctive flowers, the potato family is one of the most well-known and identifiable of all the plant families. Known variously as the tomato family or the nightshade family, *Solanaceae* contains many familiar herbaceous plants, as well as a few shrubby species.

Size

With 91 genera and 2,450 species, this is by no means a small family. The eight largest genera are *Solanum, Lycianthes, Cestrum, Nolana, Physalis, Lycium, Nicotiana,* and *Brunfelsia*. These genera combined equal almost two-thirds of all the species. One-third of the species belong to *Solanum* alone.

Solanum tuberosum,
potato

Range

Members of *Solanaceae* can be found occupying a diverse range of habitats worldwide, throughout tropical and temperate regions. The origins of the garden tomato (*Solanum lycopersicum*) are in the South American rain forest.

Origin

Like other families in the order *Solanales*, *Solanaceae* has only a scattered fossil record. However, its concentration in South America suggests that the family originated on that continent, and a very recent discovery in Argentina of a tomatillo fossil dates this family further back than expected, possibly to the Cretaceous (about 100 million years ago).

Flowers

Quite distinctive with their five fused petals and five prominent yellow anthers held together in a point at the center of the flower, it is hard to confuse them with any other family. *Petunia*, with its funnel-shaped flowers, shouldn't be mistaken for a member of the bindweed family (*Convolvulaceae*—see pages 184–185). The asymmetrical flowers of butterfly flower (*Schizanthus*) are not typical.

The fused petals give rise to a number of flower shapes. They may be round and flat, (*Solanum*), bell-shaped (*Nicandra*), or tubular (*Nicotiana*). Another fetching example is the impressive "angel's trumpets" of *Brugmansia*. There are five sepals, often partly united, and the ovary is in a superior position.

Fruit

The fruit is usually a berry, such as the familiar tomato and sweet pepper (*Capsicum*), but is sometimes a capsule, as seen in *Petunia* and *Salpiglossis*. Cape gooseberry fruit (*Physalis*) are enclosed within a papery husk. The berrylike fruits of potato plants resemble green tomatoes, but are poisonous.

Leaves

The leaves in this family are usually alternately arranged along the stem and vary greatly in size and shape. The leaves of tomato and potato, for example, are pinnately divided, while those of eggplant (*Solanum melongena*) and chili pepper are simple and entire.

USES FOR THIS FAMILY

Solanaceae must be one of humankind's most important plant families; it contains many widely grown fruits and vegetables, such as potatoes, tomatoes, sweet peppers, chili peppers, and eggplants. It also gives us tobacco (from *Nicotiana*) and many ornamentals, such as *Petunia*, *Schizanthus*, *Salpiglossis*, Chilean potato tree (*Solanum crispum*), and angel's trumpets (*Brugmansia*).

Datura metel, horn of plenty

Solanum melongena, eggplant

Flower detail showing petal and stamens

Berry in section

Poisonous

Many species in the potato family contain poisonous alkaloids. The most notorious members are deadly nightshade (*Atropa belladonna*), mandrake (*Mandragora officinarum*), Jimson weed (*Datura stramonium*), and black henbane (*Hyoscyamus niger*).

Atropa belladonna, deadly nightshade

Apocynaceae
THE MILKWEED FAMILY

This largely tropical family gets its name from the white latex released when stems or leaves are crushed. It includes trees, shrubs, herbs, succulents, and many vines, plus useful garden plants, such as milkweed (*Asclepias*), periwinkle (*Vinca, Catharanthus*), and star jasmine (*Trachelospermum*).

Size

With 4,700 species, this major family is an important component of tropical landscapes, but there are few commercially significant plants.

***Asclepias curassavica*,**
tropical milkweed
This milkweed has flowers with orange petals and protruding yellow coronas. Once fertilized, the elongated fruiting capsules form, usually in pairs, releasing numerous seeds with silken parachutes.

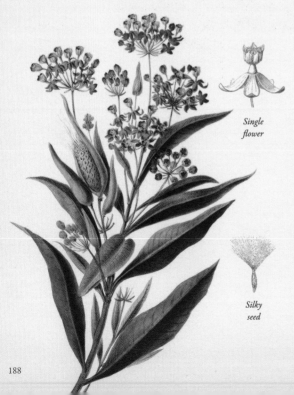

Single flower

Silky seed

Range

This family is found in all tropical regions, with many succulents adapted to arid habitats. *Apocynaceae* do reach into temperate areas, though the farther north you go, the diversity of species decreases.

Origins

Recent studies date this family at least as far back as 60 million years, with the milkweeds (formerly their own family, *Asclepiadaceae*) diverging around 40 million years ago.

USES FOR THIS FAMILY

Houseplant enthusiasts will no doubt be familiar with stalwarts such as sweet-scented wax vines (*Hoya carnosa, Stephanotis floribunda*) and rosary vine (*Ceropegia woodii*), while specialist succulent growers collect the many fetid carrion flowers (*Stapelia*) and spiny, treelike *Pachypodium*. For outside, clothe pergolas with vining star jasmine (*Trachelospermum jasminoides*) and *Dregea sinensis*; both have scented flowers. The various periwinkles (*Vinca*) make useful ground cover, while herbaceous perennial bluestars (*Amsonia*) provide delicate spring blooms and foliage color in the fall.

Trachelospermum jasminoides,
star jasmine
With fragrant blooms and evergreen foliage, star jasmine
is a valuable vine in the garden. Grow it over arbors or near
the patio to enjoy its heady perfume up close.

Leaves

The simple leaves are opposite or whorled (rarely
alternate), evergreen or deciduous, with smooth
margins. Petioles are present and stipules are
reduced or absent. Many succulents have no
leaves at all.

Pollination

Plants compete with one another for a limited
pool of visiting insects in the cutthroat world
of pollination. Some succulent species avoid the
usual butterflies and bees and target carrion flies
with blooms that look and smell like a rotting
carcass (*Stapelia*). Parachute flowers (*Ceropegia*)
trap flies in their tubular blooms, releasing them
only once loaded with a cargo of pollen.

Flowers

This family shows great diversity in pollination
strategy and, as in orchids, flowers are often highly
specialized and adapted to their individual
pollinator. The basic bloom has five sepals, five
petals, and five stamens. The sepals are partially
fused, with distinct lobes, while the petals are
fused into a tube with distinct lobes. Commonly,
the petal lobes resemble a whirling propeller,
though they can also be star- or bell-shaped. The
stamens are partially fused to the petal tube and
the anthers are free or fused into a ring surrounding
the stigma. The sticky pollen sometimes coalesces
into masses called pollinia. In many species an
outgrowth of the stamens forms a five-lobed
corona that is colorful and petal-like. Flowers can
have a sweet fragrance (*Hoya, Trachelospermum*)
or smell putrid (*Stapelia*).

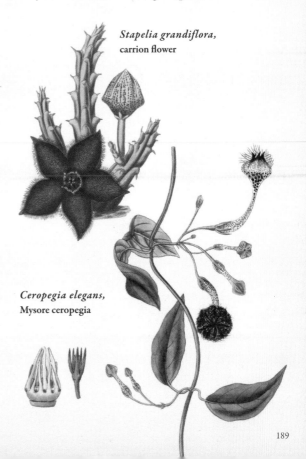

Stapelia grandiflora,
carrion flower

Ceropegia elegans,
Mysore ceropegia

Gentianaceae
THE GENTIAN FAMILY

This family is best known for the eponymous small plants with azure blue flowers, and consists mostly of herbaceous annuals and perennials. A few unusual growth forms exist, such as succulent urn gentians (*Gentiana urnula*) and ghost plants (*Voyria*), which lack chlorophyll.

Size

There are 85 genera and 1,600 species, approximately 400 of which belong to the large genus *Gentiana* (true gentians). The other major genera are *Gentianella*, *Sebaea,* and *Swertia*, all with about 100 species each.

Range

Members of *Gentianaceae* are found worldwide, and while many species are subarctic and alpine herbs, quite a few are found in salty or marshy areas. Some species, such as *Voyria* (from tropical America and West Africa) live on decaying vegetation, receiving their nutrients by parasitizing fungi.

Gentiana verna,
spring gentian

Most tropical *Gentianaceae* are shrubs or small trees, such as genus *Anthocleista*, from Africa and Madagascar. The true gentians are centered mostly around temperate regions of Asia, Europe, and the Americas. Just 13 genera are native to North America.

Origin

It is believed that the early evolutionary history of this family took place in the American tropics, during the Eocene (50–30 million years ago). However, new thinking suggests that the family may be much older, possibly dating back 100 million years.

Gentiana acaulis,
stemless gentian
Named for its trumpet-shaped azure blue flowers, which have very short stalks. The leaves form as a basal clump or in opposite pairs on short stems.

Flowers

The trumpet- or bell-shaped flowers of *Gentianaceae* are quite distinctive, although they are not always as showy as genus *Gentiana*. Colors vary from intense blue to white, cream, yellow, and even red.

The floral parts are in fours or fives, with the sepals free and the petals fused into a tube with four or five petal lobes. The stamens are attached to the corolla tube in an alternate arrangement with the lobes. The flowers of *Gentiana* have curious folds between the lobes, known as plicae.

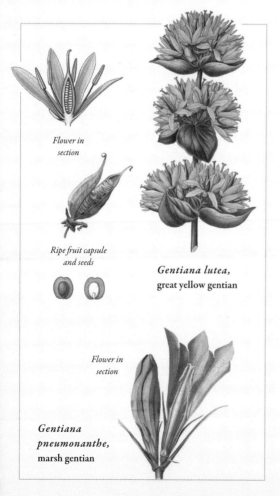

Flower in section

Ripe fruit capsule and seeds

Gentiana lutea, great yellow gentian

Flower in section

Gentiana pneumonanthe, marsh gentian

Leaves

The leaves are mostly opposite, but some species of *Frasera* have leaves in whorls of three or four, and in *Swertia* they are alternate. Many alpine or subarctic *Gentiana* bear their leaves in basal rosettes. The leaves themselves are simple in outline.

Swertia perennis, marsh felwort

Lamiaceae
THE MINT FAMILY

While perhaps best known for its aromatic perennial and annual herbs, the mint family has been expanded recently to include trees such as teak (*Tectona*), shrubs including beautyberry (*Callicarpa*), and a few vines, such as glorybower (*Clerodendrum*).

Size

The herb garden is dominated by this family, including shrubby rosemary (*Rosmarinus*), lavender (*Lavandula*), sage (*Salvia*), thyme (*Thymus*), and hyssop (*Hyssopus*), not to mention herbaceous mint (*Mentha*), oregano (*Origanum*), catmint (*Nepeta*), and basil (*Ocimum*). But within the 6,500 species of *Lamiaceae*, there are many other excellent garden plants including bee-balm (*Monarda*), dead-nettle (*Lamium*), bugleweed (*Ajuga*), and lamb's ears (*Stachys*).

Callicarpa dichotoma beautyberry

Tectona grandis, teak

Range

With a nearly worldwide distribution, only the polar regions lack *Lamiaceae*. They are especially abundant in regions with hot, dry climates, such as the Mediterranean, where the evaporation of their characteristic essential oils reduces water loss.

Origins

The mint family is one of several for which there is no fossil evidence prior to the Eocene (around 46 million years ago). With their advanced floral structures, developed to entice pollinators, it seems likely that they are one of the more recently evolved plant families.

Flowers

The shape of the flowers is perhaps the most distinctive characteristic of this family and the reason that historically they bore the name *Labiatae*, which means lipped. Each bloom has five green sepals, partially fused into a tube, from within which the five fused petals emerge. The floral tube opens out to form two lips, the upper with two lobes, the lower with three. This arrangement isn't fixed; in coleus the upper lip has one lobe and the lower lip has four lobes, while in germander (*Teucrium*) there is no upper lip and the lower has five lobes. Despite this variety, the flowers of *Lamiaceae* are readily identifiable: each usually has four stamens (two in sage and rosemary), and the petals are typically colorful and usually lack scent, though this is more than made up for by the foliage.

Teucreum botrys,
cut-leaved germander

*Single flower
from the front*

Calyx

*Single flower
from the side*

Ajuga genevensis,
blue bugle

*Flower in
section*

Rosmarinus officinalis,
rosemary

*Flower in
section*

Single flower

*Ovary in
section*

Fruit

Unlike the blooms, the fruits of *Lamiaceae* are generally unremarkable. Most species produce dry, seedlike nutlets and these remain concealed within the sepal tube after the petals drop. There are typically four nutlets per flower and they fall to the ground when dislodged. Some species have developed fleshy fruits, which are consumed and dispersed by animals. The harlequin glorybower (*Clerodendrum trichotomum*) has four-lobed blue fruits set in the center of distinctive red sepals, making an attractive target for hungry birds.

Leaves

When identifying plants, the combination of opposite leaves and square stems immediately brings *Lamiaceae* to mind. It is a useful character combination and easy to spot, but be aware that it is not entirely unique. *Lamiaceae* leaves are usually simple, more rarely lobed or compound, and often fragrant when bruised. Most species have an opposite arrangement, with each pair of leaves at right angles to the previous pair, though leaves are occasionally whorled. Drought-resistant species, such as rosemary, often have thick leathery leaves, or as in *Stachys byzantina*, the leaf surface is covered in silver hairs to reduce evaporative water loss. Conversely, plants from humid (basil) or aquatic habitats (mint) have thin leaves with few or no hairs.

Petal tube opened

Single flower

Calyx tube opened

Four nutlets

Salvia officinalis,
common sage
The dry fruits of sage, known as nutlets, remain concealed within the calyx, eventually falling out. Though actually fruits, they look and behave like seeds.

Clerodendrum trichotomum,
harlequin glorybower

Mentha longifolia,
brook mint
This widespread mint can be found from western Europe through to China and south to South Africa. Like many *Mentha* species, it has a strong menthol fragrance.

Pollination

More than 900 species make *Salvia* the largest genus in the mint family. The culinary sage (*Salvia officinalis*) is just one of many useful garden forms, but why are there so many species? The answer may lie in the unique pollination mechanism within the flower. Each bloom has only two stamens, which are elongated to form levers. When a pollinator visits a bloom seeking nectar, it pushes the lever, forcing the pollen-packed anthers down onto the pollinator's back. Slight changes in stamen length will cause the pollen to be deposited on a different part of the pollinator's body, meaning it won't be picked up by the fertile stigma of another flower. Thus, a barrier is created between previously inter-fertile plants, and a new *Salvia* species is born.

While an essential element of the edible garden, *Lamiaceae* also contains useful decorative plants for many situations. Gardeners seeking to reduce water usage would do well to plant shrubs such as tree germander (*Teucrium fruticans*), Jerusalem sage (*Phlomis fruticosa*), *Caryopteris*, and chaste tree (*Vitex agnus-castus*), which can withstand considerable drought. Conversely, where water is not in short supply, various ornamental basils, sages, and coleus (*Solenostemon scutellarioides*) make great summer container plants. For low-growing hedges to line borders, rosemary, hyssop, lavender, and wall germander (*Teucrium chamaedrys*) are ideal.

Vitex agnus-castus,
chaste tree

Solenostemon scutellarioides,
coleus

Oleaceae

THE OLIVE FAMILY

Another family with a number of suprising relatives, *Oleaceae* is home to olive and ash trees, as well as the shrubs lilac (*Syringa*), jasmine (*Jasminum*), privet (*Ligustrum*), *Osmanthus*, *Forsythia*, and *Abeliophyllum*. All family members are woody trees and shrubs, with a few scrambling climbers.

Size

This medium-size family presently contains 24 genera and 800 species. Uncertainty over the number of species is largely down to the genus *Jasminum* (jasmine), which may contain between 200 and 450 species, depending on which authority you consult.

Jasminum didymum,
desert jasmine

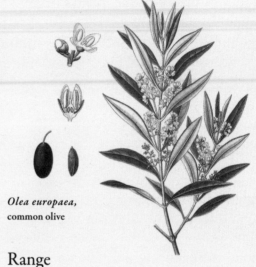

Olea europaea,
common olive

Range

The olive family is represented widely throughout the world, from tropical to subarctic regions. Ash trees (*Fraxinus*) are found across most of the temperate northern hemisphere; *Jasminum* are seen across Europe, Asia, Australia, the Pacific, and tropical America. *Ligustrum* is found across Europe to southeastern Asia and *Abeliophyllum* is localized to Korea.

Origin

Even though the fossil record of the olive family is quite sparse, botanists agree that its origins stretch back to the Mesozoic. Some genera that presently show a localized distribution are likely to have been isolated for many millions of years.

Flowers

Oleaceae flowers are generally quite small and often overlooked, perhaps appreciated more for their fragrance or fruit than their individual looks. Some species, such as *Fraxinus* and *Ligustrum*, are not valued for their flowers. It is for this reason that amateurs may fail to spot their one shared characteristic: floral anatomy.

The flowers have four united sepals, four united petals, and two stamens, which is a relatively unusual combination. They are regular in shape, produced in large numbers, and are often scented. The ovary is in a superior position and has two united carpels, which ripen into a two-chambered fruit.

Fraxinus is a bit of an oddity because the flowers of many species in this genus are without petals. The number of petals in *Jasminum* varies from four to nine. Often with species in *Oleaceae*, all the petals are fused into long or short tubes.

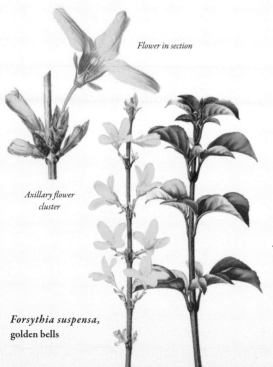

Flower in section

Axillary flower cluster

Forsythia suspensa,
golden bells

Leaves

The leaves in *Oleaceae* are oppositely arranged. Some species are evergreen.

Fruit

Jasminum and *Ligustrum* form berries. *Forsythia* makes dry capsules containing winged seeds. The fruit of *Olea* (olives) are small drupes. *Fraxinus* produce winged fruit called samaras, which are commonly referred to collectively as "keys."

Scrophulariaceae

THE FIGWORT FAMILY

Few plant families have undergone as much taxonomic upheaval as this one, with many species removed into other families (mainly *Plantaginaceae* and *Orobanchaceae*) or brought in (*Buddleja* from *Loganiaceae*). It now includes trees, shrubs, and both annual and perennial herbs.

Size

Now reduced to around 1,800 species, *Scrophulariaceae* contains yard-worthy plants, such as figwort (*Scrophularia*), mullein (*Verbascum*), Cape fuchsia (*Phygelius*), *Nemesia*, and *Diascia*. Most are herbs, though shrubs and trees can be found in butterfly bush (*Buddleja*) and elsewhere.

Range

Widely distributed in tropical and temperate areas, the figworts are especially common in southern Africa and at high elevation on tropical mountains.

Origins

This family is unknown in the fossil record, though its order, *Lamiales*, which also includes *Plantaginaceae*, *Lamiaceae*, and *Oleaceae*, has been recorded in the Eocene.

Flowers

Most blooms are arranged in elongated inflorescences (solitary flowers do occur) and the flowers can be bilaterally symmetrical, providing enlarged landing platforms for bees, or radially symmetrical (*Buddleja*, *Verbascum*). The four or five sepals are free or fused, while the four or five petals are always fused and usually tubular. As the tube opens out, the lobes can be of different sizes, forming two lips, or evenly sized, while some have tubular spurs behind the blooms. Stamens are partially fused to the petal tube and come in a pair or two pairs (one short, one long). Some *Scrophulariaceae* have scented flowers, including night phlox (*Zaluzianskya*) and several *Buddleja*.

Buddleja davidii,
butterfly bush

Diascia aliciae,
Alicia's twinspur

Flower face

Spurs on rear

Tubular flowers

*Zaluzianskya
maritima,*
night phlox

The all too familiar butterfly bush (*Buddleja davidii*), while an excellent source of food for pollinating insects, does have a reputation for escaping the confines of domestic gardens. For something different, try growing *B. colvilei* against a wall; its pink flowers are much larger than those of the regular butterfly bush. The fiery orange, ball-like inflorescences of large shrub *B. globosa* make a striking display against the dark green foliage. If you're a fan of the unusual you might wish to locate chocolate-scented perennial *Glumicalyx goseloides*. Its pendulous flowers appear white, but open to reveal zesty orange petals; plant it in full sun on well-drained soil. Finally, the Cape fuchsias are small shrubs or herbs dripping with tubular flowers; peer inside the cylinder of petals to admire the contrasting colors within.

Buddleja colvilei,
Colvile's
butterfly bush

Leaves

All family members have simple leaves (though they can be opposite, alternate, or whorled) and with entire or toothed margins. Petioles are present or absent, and stipules are usually absent (sometimes present in *Buddleja*). A few *Scrophulariaceae*, such as *Scrophularia* itself, have square stems and opposite leaves, thus resembling the mint family (*Lamiaceae*—see pages 192–195), but the latter are readily distinguished by their scented foliage.

Scrophularia vernalis,
yellow figwort

Plantaginaceae
THE PLANTAIN FAMILY

While a handful of trees and shrubs are included in this family, most species are annual, biennial, or perennial herbs. Few are cultivated as crops, though there are many ornamentals such as foxglove (*Digitalis*), snapdragon (*Antirrhinum*), toadflax (*Linaria*), and beardtongue (*Penstemon*).

Size

Previously consisting of 250 species, including the common lawn weed plantain (*Plantago*), this family has expanded to about 1,900 species. Most of these are terrestrial herbs, though there are a few partly woody climbers (*Asarina*, *Rhodochiton*) and aquatics (*Callitriche*, *Hippuris*).

Digitalis purpurea,
foxglove

Range

This family is found worldwide except in Antarctica, though it is most often associated with temperate regions.

Origins

As with the closely related *Scrophulariaceae*, little is known of this family's history.

Flowers

In most respects, the flowers of *Plantaginaceae* resemble and are difficult to distinguish from those of *Scrophulariaceae*. Each bloom has four or five fused sepals and four or five fused petals, though the latter are absent in *Plantago*. In *Antirrhinum*, *Linaria*, and others, the petals are shaped into two opposing lips, with the lower often puckered upward and forming the characteristic snapdragon mouth. There are four stamens, two short and two long, though only two in *Veronica*. Many *Penstemon* have a fifth, infertile stamen covered in hairs, which is the source of the name "beardtongue."

Bearded tongue

Penstemon richardsonii,
Richardson's beardtongue

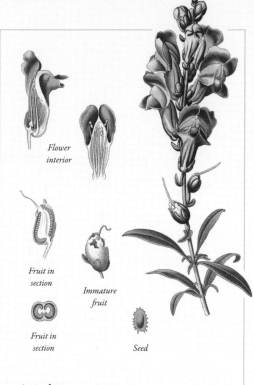

Flower interior

Fruit in section

Immature fruit

Fruit in section

Seed

Antirrhinum majus,
snapdragon

Snapdragon petals are marked with yellow, a color readily seen by bees, showing them where to land. Only heavy bumblebees can prise open the lobes to reach the nectar within.

Without *Plantaginaceae* we would lose many fine border plants. The stately spires of foxglove and *Veronicastrum* form a useful backdrop, while *Penstemon* and turtlehead (*Chelone*) fill the middle ground. Throw in some annual snapdragons in the foreground, and plug any gaps with obelisks covered in tender climbers, such as *Lophospermum* and *Rhodochiton*, and your border is built. The more minuscule members of the family perform well in alpine gardens and troughs. Carpet-forming *Globularia* has blue flowers and mixes well with pink fairy foxglove (*Erinus alpinus*). The diminutive *Veronica* (formerly *Hebe*) *epacridea* has dainty leaves neatly arranged along wiry stems and would make a good foil for purple *Penstemon davidsonii*. If you think that purple, pink, blue, and white shades seem to predominate in *Plantaginaceae*, the fiery flowers of *Penstemon pinifolius* are notable exceptions, available in outrageous orange or sulfurous yellow.

Globularia alypum,
Alypo globe daisy

Leaves

Opposite leaves are most common, though alternate (*Digitalis, Globularia*) and whorled (*Hippuris*) arrangements can be found. Within the large genus *Veronica*, which now includes *Hebe*, there are species with alternate leaves and those with opposite leaves. Leaf margins are entire or toothed, petioles are present or absent, and stipules are absent.

Flowers lack petals

Leaves in whorls

Hippuris vulgaris,
bottlebrush

One of a handful of aquatic plants in the plantain family, bottlebrush or mare's tail has greatly reduced, petal-less flowers.

Boraginaceae
THE BORAGE FAMILY

The uncoiling inflorescences made up of pretty flowers in shades of blue make this family of rough-textured herbs easily recognizable. While there are some shrubby or treelike species, such as manjacks (from genus *Cordia*), most family members are low annuals and perennials that die back after flowering.

Size

There are 142 genera covering 2,450 species. Despite being named for genus *Borago*, there are only five species of borage; the larger genera include *Cordia* and heliotrope (*Heliotropium*), both of which have about 300 species each.

Range

Members of the borage family are located throughout temperate and subtropical regions of the world. They are found less frequently in cool temperate and tropical areas. The major center of diversity is around the Mediterranean Basin.

Cordia myxa,
Assyrian plum

Borago officinalis,
borage

Although the woody genus *Cordia* can be found around the world, it is mostly limited to warmer regions, such as Africa and southeastern Asia. About one-fifth of all genera are native to North America.

Origin

There is a rich fossil record of the *Boraginaceae* from Miocene and Pliocene deposits in North America. Its abundance and diversity suggests that the family evolved during the Cenozoic Era (approximately 50 million years ago).

Flowers

Individually, the flowers are five-parted, with five sepals, five petals, five stamens, and a superior ovary. The petals are fused into a five-lobed corolla, forming a tube or bell. The sepals alternate with the corolla lobes; in *Borago*, this gives the flowers a starlike appearance.

The petals are typically blue, but other common colors are yellow, white, pink, and purple. Sometimes the flowers change color with age, possibly as an indicator to insect pollinators. Usually the flowers are regular in shape, but they are lipped in *Echium* and related genera.

The flower heads are described as scorpioid or helicoid, meaning that they are coiled like a spring or scorpion's tail. These unfurl gradually as the flowers open, which is a recognizable trait in this family. The tall spires of *Echium* are unmistakable and impressive.

Echium strictum,
Narrow viper's bugloss

USES FOR THIS FAMILY

Ornamental species include *Heliotropium*, bluebell (*Mertensia*), forget-me-not (*Myosotis*), lungwort (*Pulmonaria*), *Echium*, *Brunnera*, and the pretty annual *Cerinthe*. Borage (*Borago officinalis*) is an annual garden herb; it is also an excellent source of nectar for bees. Comfrey (*Symphytum*) is useful to organic gardeners as a compost activator and a source of fertilizer.

Myosotis scorpioides,
water forget-me-not

Leaves

These are coarse-textured plants with rough hairs covering the leaves, stems, and flower heads. Contact with the skin can cause an itchy rash. The alternately arranged leaves are typically narrow and simple in shape, with smooth or toothed margins. Sometimes the leaves are patterned, as in genus *Pulmonaria*.

Asteraceae
THE DAISY FAMILY

The most familiar members of this family are herbaceous perennials, such as aster, coneflower (*Echinacea*), and black-eyed Susan (*Rudbeckia*), or annuals, including sunflower (*Helianthus*) and cornflower (*Centaurea*). However, this mega-family also includes trees, shrubs, vines, alpines, and many important garden weeds and wildflowers, such as dandelions, daisies, and thistles.

Size

With around 23,600 species, the *Asteraceae* is the largest plant family on Earth. Given its size, you might expect it to contain numerous important edible plants, but sunflower, lettuce, artichokes, and chicory are the only significant crops. It's in the garden where this family proves its worth and the list of useful perennials is vast; two genera are the focus of great devotion from exhibitors, *Dahlia* and *Chrysanthemum*.

Achillea millefolium,
yarrow

Centaurea cyanus,
cornflower

Range

Absent only from Antarctica, the daisy family can be found in almost every ecosystem, though they're generally rare in tropical rain forest. They are especially diverse in arid regions, including deserts and areas with a Mediterranean-type climate, and at high elevation on tropical mountains.

Origins

Perhaps one of the most advanced of all plant families, the earliest recording of the *Asteraceae* is in Cretaceous deposits in Antarctica (76–66 million years ago). Further confirmation of their southern hemisphere origins can be found in their nearest relation, the small family *Calyceraceae*, which is entirely restricted to South America.

Flowers

The daisy family was known historically as the *Compositae*, because what appeared to be individual flowers were clusters of tiny blooms fused together into a head (i.e. the "flower" was a composite of several flowers). These heads, technically known as capitulae, perform the same task as a single flower, attracting pollinating insects. The tiny flowers (or florets) within the heads usually retain in miniature all the major flower parts. The sepals are reduced to a cluster of scales or hairs (the pappus), while the five petals are fused into a tube with the five stamens partially attached.

Many daisies have two types of floret and these are distinguished by their petals. Disc florets have a symmetrical petal tube, while ray florets are asymmetrical, with the petal tube opening to form a lip on one side. In the common daisy (*Bellis perennis*), the disc florets are yellow and in the center of the head, while the white rays line the outside. Dandelions (*Taraxacum*) have only ray florets, while groundsel (*Senecio vulgaris*) has only disc florets. Surrounding the whole head, and reminiscent of sepals, is a ring of green bracts, known as the involucre. The composite inflorescence can be found in all *Asteraceae*.

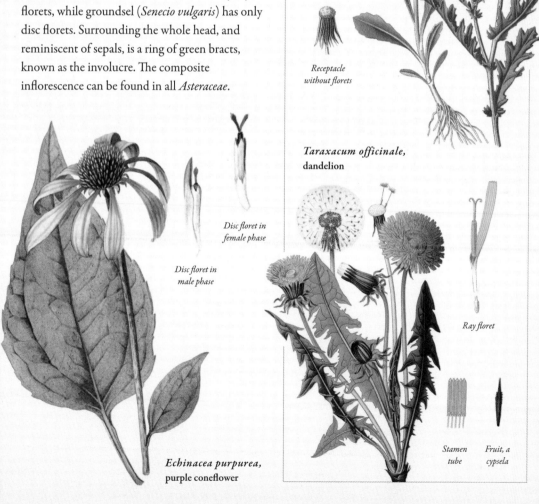

Senecio vulgaris,
groundsel

Disc floret

Capitulum   In section

Fruit, a
cypsela

Receptacle
without florets

Taraxacum officinale,
dandelion

Disc floret in
female phase

Disc floret in
male phase

Ray floret

Stamen
tube

Fruit, a
cypsela

Echinacea purpurea,
purple coneflower

Fig.1.1

Fig.1.1. *Bellis perennis*, common daisy. The "flower" of *Asteraceae* is actually a greatly reduced inflorescence called a capitulum. The receptacle (left) bears tiny flowers or florets that have asymmetrical (middle) or symmetrical (right) petal tubes.

Fig.1.2. The "seeds" of *Helianthus annuus* (sunflower) are a type of fruit called a cypsela; the real seed is inside the striped shell.

Fig.1.2

Fruit

Given the small size of the actual flowers in daisy heads, unsurprisingly the fruits are also very small. Not all florets in each head are fertile, but those that are produce one fruit each—a dry fruit known as a cypsela. In many species, the hairy or toothed pappus remains attached to the fruit and provides a means of dispersal, by becoming attached to a passing animal or by acting as a wing. In sunflowers (*Helianthus*), the characteristic black-and-white striped "seed" is a fruit, with the real seed enclosed within the dry outer shell; the pappus falls off before dispersal.

Leaves

As befits such a large, diverse family, daisy leaves come in many shapes and sizes. They can be entire or lobed, with or without petioles, and several exude white sap when damaged. Dahlias have more complex compound leaves, which can

Carduus crispus,
welted thistle

Dahlia imperialis,
tree dahlia
One of the giants of the
daisy family, tree dahlia
can reach 33 feet tall
but is an herbaceous
perennial.

Guizotia abyssinica,
nyger seed plant

While relatively few daisies are grown
commercially, many contain chemicals that
deter hungry herbivores. Some of these plants
are also useful in the garden as herbs or insect
deterrents. Garden pyrethrum (*Tanacetum
coccineum*) is the source of an insecticide, also
called pyrethrum, which can be extracted from
the dried flower heads. Planting it as a companion
to vulnerable crops will also partially protect
them from pests. Mexican marigolds (*Tagetes*)
deter a variety of crop pests, including onion fly
and whiteflies, while chamomile (*Matricaria
chamomilla*) improves crop flavors and attracts
predatory hoverflies. Edible herbs include curry
plant (*Helichrysum italicum*) and tarragon
(*Artemisia dracunculus*).

With so many species available, there is a daisy
in flower at almost every time of year. Their peak,
however, is in late summer and early fall, when
fading light and morning dew highlight their
distinctive heads. Inject color into fading summer
borders by adding asters, coneflowers, yarrow
(*Achillea*), heleniums, or exotic dahlias. And after
the first frost, the seed heads of globe thistles
(*Echinops*) and cardoons (*Cardunculus*), with
their architectural forms, provide natural
sculpture throughout the dark days of winter.

exceed 3 feet in length in tree dahlia (*Dahlia
imperialis*). Leaves are often arranged alternately
along the stems, but can also be opposite or
whorled, and herbaceous genera typically have
a rosette of leaves at the base.

Wildlife magnets

Many garden daisies provide seeds for hungry
birds; flocks of finches are drawn to old sunflowers
to feast on their seeds, or to feeders filled with
nyjer, harvested from another daisy (*Guizotia
abyssinica*). In summer, too, the daisies are wildlife
magnets, drawing in bees, hoverflies, and
butterflies to their nectar-rich blooms.

*Tanacetum
coccineum,*
painted daisy

Campanulaceae
THE BELLFLOWER FAMILY

The bellflower family now contains lobelias, making this a much larger group of mostly herbaceous perennials (some are annual and biennial) with beautiful bell-shaped or lipped flowers that are predominantly in shades of blue. A few species are shrubby, and giant lobelias grow in a manner that resembles small palm trees.

Size

The inclusion of the former lobelia family (with about 30 genera) practically doubles the size of *Campanulaceae*, which now has 79 genera and 1,900 species. Most of the species are split across two major subfamilies, which correspond more or less to *Campanulaceae* and *Lobeliaceae* families before they were merged.

Range

The vast majority of *Campanulaceae* come from cool temperate regions, but this family extends across the globe. The *Lobeliaceae* side of the family tends to have a warm-temperate and tropical distribution as compared to the cool-climate *Campanulaceae*. There is a scattering of genera across the southern hemisphere, with a well-defined population of several genera in South Africa.

Origin

Although the fossil pollen record dates the origin of this family to 30 million years ago, its global diversity suggests a much earlier origin in the Late Cretaceous (80–70 million years ago). The *Campanulaceae* and *Asteraceae* (daisy family) are closely related, sharing a common ancestor.

USES FOR THIS FAMILY

Species from *Lobelia* and almost all from *Campanula* are easy-to-use plants for the garden, and low maintenance. A few others, such as *Symphyandra*, *Phyteuma*, *Edraianthus*, and *Jasione* are also grown. A number of species, such as those in *Adenophora*, *Codonopsis*, and *Platycodon*, are valued in rock gardens. *Campanula rotundifolia* is the Scottish bluebell (also known as harebell), quite a different plant from the English bluebell, and does well in rock gardens or on sunny rocky banks.

Roella ciliata,
South African harebell

*Open flower in
section, pollen shed,
stamens withered*

Campanula rotundifolia,
common harebell

*Flower in
section, before
it opens*

Lobelia cardinalis,
cardinal flower

*Flower in
section*

*Withered flower with
developing fruit*

*Fruit capsule
in section*

Flowers

The prominent and often delicate flowers of the *Campanulaceae* tend to fall into one of two camps, depending on which side of the family they come from. The flowers of subfamily *Campanuloideae* are typically bell-shaped with five similar petals. Three of the five petals of subfamily *Lobelioideae* are enlarged to form a lower lip.

The petals are usually united to create a cup-, bell-, or tube-shaped corolla, and the calyx is united with the ovary. The corolla is inserted where the calyx becomes free from the ovary. The petals are not united in the genera *Jasione, Asyneuma, Michauxia, Cephalostigma,* and *Lightfootia.*

Flowers are usually carried in racemose or cymose flower heads, but sometimes singly. Sheep's bit scabious (*Jasione montana*), round-headed rampion (*Phyteuma orbiculare*), and some lobelias have flowers densely packed into tight flower heads.

Many species are valued for their large colorful flowers, which are usually blue, but can also be red, purple, or white, and sometimes yellow. Bees are frequent visitors, particularly to blue flowers, but there are many other insect pollinators. Red-flowered species can be visited by butterflies or birds, such as the ruby-throated hummingbird, which pollinates the spectacular cardinal flower (*Lobelia cardinalis*) in North America.

*Legousia
speculum-veneris,*
Venus's looking glass

Apiaceae
THE CARROT FAMILY

The recognizable flat-topped herbs of *Apiaceae* make it one of the best-known plant families. The botanist John Ray, in the 16th century, was the first to recognize this family and its shared characteristics with the monocots. These species are mostly upright and branched, nonwoody annuals and perennials with hollow stems, commonly known as umbellifers.

Size

This is one of the larger flowering plant families with more than 400 genera and 3,500 species. Most genera only contain a handful of species, and sometimes just two or three; the genus celery (*Apium*), after which the family is named, is a fairly typical example, with just 20 species.

Range

Family members are present in most parts of the world, with the greatest concentration in temperate areas, and two-thirds of all species native to Europe and Asia. *Drusa glandulosa* from the Canary Islands and *Naufraga balearica* from the Mediterranean are both oddities because their closest allies are, inexplicably, in South America.

Origin

Apiaceae's wide pattern of distribution reflects a long history of evolution. There is a close affinity with the aralia family (*Araliaceae*—see pages 212–213), and it is strongly believed that the two families share a common ancestry, which is believed to extend back to the Late Cretaceous (80–70 million years ago).

Pastinaca sativa,
parsnip

Flowers

The flat-topped umbellate flower head is the key feature of this family, hence its former name, *Umbelliferae*. In an umbel, the individual flower stalks radiate from a single point, like the spokes of an umbrella. Compound umbels are frequently seen, whereby the umbels themselves are arranged

Ferula assa-foetida,
asafoetida

Foeniculum vulgare,
fennel

Daucus carota,
carrot

into smaller umbels, called umbellules. One exception is sea holly (*Eryngium*), with its dome-shaped, stalkless umbels.

There can be enlarged bracts around the whole umbel or individual umbellules, as seen in various species of *Eryngium* and hare's ear (*Bupleurum*). An extreme example is the Mexican *Mathiasella bupleuroides*, which bears a slight resemblance to *Helleborus*, from *Ranunculaceae* (see pages 106–109).

The tiny individual flowers are five-parted with a very reduced or absent calyx. They come in a wide range of colors; from green to white, yellow, pink, and purple. *Eryngium* are once again the exception because they feature steely blue flowers. The outer flowers in the umbel are sometimes irregular, as in carrot (*Daucus carota*). This may serve to attract insect pollinators, which include flies, mosquitoes, bees, butterflies, and moths.

Leaves

Umbellifer leaves are alternately arranged and usually deeply lobed or divided, sometimes pinnate or fernlike. They vary in size, and when crushed they often release a strong aroma.

Araliaceae

THE ARALIA FAMILY

The better-known members of *Araliaceae* are plants of damp forest and woodland habitats, such as ivies, aralias, and fatsias. Some are trees and shrubs, such as angelica tree (*Aralia elata*). Others are climbing species, such as those in genus *Hedera* (ivy). The family also includes small herbaceous species, for example those belonging to *Panax* (ginseng).

Size

This is a medium-size family of 39 genera and 1,425 species. The woody genera *Schefflera* (700 species) and *Polyscias* (116 species) are two of the largest groups, compared to *Aralia* (68 species) and *Panax* (11 species). There are about 12 species of ivy (*Hedera*).

Range

The species of *Araliaceae* are found all over the world, although it is chiefly a subtropical and tropical family, with the main centers of diversity in southern and southeastern Asia and tropical America. Typical habitats include mountain cloud forests, rain forests, and damp, humid woodlands.

Flower in section   *Individual flower*

Aralia spinosa,
angelica tree

USES FOR THIS FAMILY

Many members of the aralia family are valued as foliage plants for shaded areas, from climbing ivies (*Hedera*) to the shrubby *Fatshedera* and *Schefflera*. The spectacular, billowy flower heads of *Aralia* and *Fatsia* should be regarded as a bonus. There are many cultivated forms with interesting leaf shapes, variegation, or patterning. Ivy nectar is a valuable winter food source for many insects.

Origin

There is an extensive fossil record for *Araliaceae* that goes back to the Late Cretaceous (80–70 million years ago). This suggests that the family originated in North America and gradually spread across the Bering Land Bridge into Asia and Europe.

Flowers

Individually, the flowers are small or tiny and regular in shape. They are usually green, pale yellow, or whitish with five petals, five stamens, and five often very small sepals. On pollination, each flower matures into a red or purple, berrylike, five-seeded drupe.

Typically, the flower heads are simple and rounded umbels, in contrast to the flat-topped and frequently compound umbels of the closely related carrot family (*Apiaceae*—see pages 210–211). Some members produce elongated panicles, such as species belonging to *Aralia*.

Leaves

Perhaps the most prominent feature of this family is the foliage. Borne alternately along the stems, the leaves are often large and interestingly lobed or split into smaller leaflets.

Fatsia japonica,
Japanese aralia

Hedera helix,
ivy

The main genera of this family can be grouped according to their distinctive leaf anatomy. *Kalopanax*, *Fatsia*, *Hedera*, and *Tetrapanax* all have lobed but simple leaves. *Aralia*, *Polyscias*, *Schefflera*, *Acanthopanax*, and *Panax* all have compound leaves; those of *Aralia* and *Polyscias* are pinnate, the other three are palmate. Juvenile leaves tend to be shaped differently from the adult leaves.

Stems

Those species with a climbing habit often exhibit two phases of growth: adult and juvenile. Aerial roots form on juvenile shoots, so that the stems can cling and climb, and the adult stems are self-supporting.

Adoxaceae

THE ELDER FAMILY

Formerly part of the honeysuckle family (*Caprifoliaceae*—see pages 216–217), the viburnums, elders (*Sambucus*), and moschatels (*Adoxa*, *Sinadoxa*) are shrubs, small trees, or perennial herbs. The woody species in *Adoxaceae* are important garden plants and several produce edible fruits. The herbs are more often weeds and wildflowers.

Size

The great majority of the 225 species in *Adoxaceae* belong in the large genus *Viburnum*. In many ways, it resembles *Hydrangea* (see pages 176–177) because both are shrubs or small trees with opposite leaves and sometimes sterile flowers. But, viburnums have fleshy fruits, while hydrangeas have dry capsules.

Range

Widely distributed in temperate Asia, Europe, and North America, *Adoxaceae* also reaches down into South America, southeast Australia, and several mountains in Africa.

Origins

Numerous leaf fossils from the late Cretaceous (99–93 million years ago) have been identified as *Viburnum*, though this seems early, given the position of *Adoxaceae* on the plant family tree (see pages 8–9). Elder fossils date from the Late Eocene (around 40 million years ago).

Flowers

In *Viburnum* and *Sambucus*, the numerous small flowers are held in flat-topped or spherical inflorescences at the stem tips. In contrast, moschatels have only a handful of flowers in each inflorescence, either on short branches (*Sinadoxa*) or clustered tightly (*Adoxa*). Most flowers have five sepals fused together and four or five petals fused at the base. The three to five stamens sit opposite the sepals and are attached to the petals. Several *Viburnum* produce showy sterile flowers around the edges of the inflorescence and, in some cultivars, only sterile flowers (snowball bush). Blooms can be sweetly fragrant (*V. carlesii*, *V. farreri*), unpleasant (*V. lantana*), or unscented.

Adoxa moschatellina,
moschatel

Single flower

Immature fruits

Mature fruits

Sambucus ebulus,
dwarf elder

Fruit

The fruits of this family are fleshy with hard pits inside and are typically red, blue, or black.

Leaves

The herbaceous moschatels have leaves in loose, basal rosettes, while woody elders and *Viburnum* show an opposite arrangement. Leaves are evergreen or deciduous, simple in *Viburnum*, but compound in the other genera, with one to three leaflets in moschatels and pinnate leaves in *Sambucus*. Leaf edges are toothed, sometimes entire, and stipules are present or absent.

USES FOR THIS FAMILY

Elders, especially *Sambucus nigra*, are grown for their fruits and flowers, which are used to flavor a variety of foods. They are popular plants with bees and birds, and several cultivars have colorful and/or lacy foliage.

The diverse genus *Viburnum* provides numerous garden-worthy plants. For hedging, it's hard to cranberrybush (*V. opulus*), with white flowers, glossy red fruits, and dazzling fall color. If you'd prefer an evergreen hedge, laurustinus (*V. tinus*) is a stalwart with its useful winter blooms. It's in winter and spring when these shrubs come into their own, with the classic *V. × bodnantense* budding up while under a blanket of snow. The blooms of *V. bitchiuense* and *V. farreri* follow, with their strong scents, which are especially welcome on a cool spring morning.

Viburnum opulus,
European cranberrybush

Sterile flowers

Fertile flowers

Caprifoliaceae
THE HONEYSUCKLE FAMILY

Though now shorn off the *Adoxaceae*, this family has grown to encompass teasels, scabious, and valerians. It comprises small trees, shrubs, and vines, plus annual, biennial, and perennial herbs. A few species are edible, such as corn salad (*Valerianella*) and honeyberry (*Lonicera caerulea*).

Size

Given their great diversity, the 900-plus species of *Caprifoliaceae* have numerous uses. Popular plants in this family include herbaceous valerian (*Valeriana, Centranthus*), pincushion flower (*Scabiosa, Knautia*), whorlflower (*Morina*), and vining honeysuckle, as well as shrubs (*Abelia, Weigela, Leycesteria*).

Lonicera
hildebrandiana,
giant honeysuckle

Range

They are widespread across the northern hemisphere and in Africa and South America. One species, the twinflower (*Linnaea borealis*), occurs across all three northern continents, while several honeysuckles (*Lonicera*) and teasels (*Dipsacus*) have become pervasive weeds.

Origins

Fossilized fruits very similar to those of modern *Dipelta* come from the Late Eocene (around 40 million years ago), while fruits resembling those of *Patrinia, Valeriana*, and *Heptacodium* have been found from later in the Miocene (11–5 million years ago).

Flowers

The basic bloom in this family has four or five fused sepals and four or five petals partially fused into a tube. The petal lobes may be symmetrical (*Heptacodium*) or with two upper and three lower lobes (*Dipelta*) or four upper and one lower lobe (some *Lonicera*). The filaments of the one to five stamens are fused to the petals.

Individual flower

Valeriana officinalis,
valerian

Symphoricarpos racemosus, snowberry

Fruit in section

USES FOR THIS FAMILY

Honeysuckle is of course best known for its scent, and none is more welcome than that produced by the blooms of early spring-flowering shrub *Lonicera fragrantissima.* The giant Burmese *L. hildebrandiana* is rather tender, but its 6 inch-long flowers are deliciously fragrant. While not all honeysuckles are scented, most abelias are, and these have other attractions, including colorful bracts and, in some cases, variegated foliage. If there's no room for a shrub or vine, herbaceous valerian (*Valeriana officinalis*) is a good choice. This hardy, self-seeding perennial has tall, airy flowering stems tipped with white blooms that exude a heady vanilla scent.

Various arrangements of flowers can be found and most include plentiful bracts in the inflorescence. Some species (*Lonicera involucrata, Heptacodium, Abelia*) have colorful bracts that provide interest long after the petals fall. In teasels and their kin, numerous small flowers are fused together into a composite head, like those of the daisy family. They are surrounded by an involucre of bracts, and the outermost flowers have much larger petals.

Fruit

Fruits are variable and can be dry (*Weigela, Abelia*) or fleshy (*Lonicera, Symphoricarpos*).

Leaves

The shrubs and vines have opposite or whorled leaves, while in the herbs, the leaves often form a rosette at the base. They are usually simple, though occasionally divided with a feather-like (pinnate) leaflet arrangement (*Knautia, Cephalaria*). Leaf margins are entire or toothed, petioles are distinct and stipules are absent.

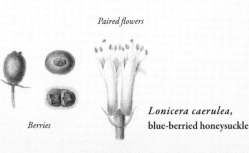

Paired flowers

Berries

Lonicera caerulea, blue-berried honeysuckle

Glossary of botanical terms

bulb: An underground storage unit composed of fleshy, scalelike leaves, as in onions.

calyx: The sepals of a flower, typically forming a whorl, that encloses the petals and forms a protective layer around a flower in bud.

capitulum: A compact headlike inflorescence, in particular a dense flat cluster of small flowers or florets, as in plants of the daisy family.

carpel: The female reproductive organ of a flower, consisting of an ovary, a stigma, and usually a style. It may occur singly or as one of a group.

compound: (of a leaf) Single leaf divided into two or more leaflets. See also *simple*.

corm: A swollen bulblike underground stem, common in the iris family, e.g. *Crocus*.

corolla: The petals of a flower, typically forming a whorl within the sepals and enclosing the reproductive organs.

corona: A ringlike structure in-between the petals and stamens. Forming the trumpet of daffodil flowers.

corymb: A flat-topped inflorescence, similar to an umbel, but the flowers are not joined at one point.

cultivar: Cultivated variety; a plant that has been produced in cultivation by selective breeding.

cupule: A cup-shaped structure formed from bracts. In the oak family, the cup in which acorns sit.

cyme: A flower cluster with a central stem bearing a single terminal flower that develops first, the other flowers in the cluster developing as terminal buds of lateral stems.

drupe: A fleshy fruit with thin skin and a central stone containing the seed, e.g. a plum, cherry, almond, or olive.

entire: (of a leaf or leaflet) An undivided leaf with no lobes or teeth along the margins.

epiphyte: A plant that grows attached to another plant, but without parasitizing it, as in many orchids and bromeliads.

eudicot: The largest group of flowering plants; characterized by branching leaf veins, flowers with parts in multiples of four or five, e.g. roses, legumes, daisies, heathers.

exudate: Fluids exuded from cut leaves or stems. May be sticky, white (spurge family), colored (poppy family), or clear.

filament: The slender part of a stamen that supports the anther.

hesperidium: A fruit with sectioned pulp inside a separable rind, e.g. orange or grapefruit.

hypanthium: A cuplike or tubular enlargement of the receptacle of a flower, loosely surrounding the carpels or united with them.

inflorescence: A cluster of flowers, which can have many arrangements, e.g. cyme, panicle, raceme, spike, etc.

intergrade: When plant organs blend into one another, as when some petals bear partially formed anthers.

involucre: The ring of sepal-like bracts that surrounds a capitulum.

keel: (of petals) The fold down the middle, or the two fused lower petals of a typical legume flower.

ligule: A leafy or hairy projection arising where the leaf joins the sheath, as in some grasses and gingers.

lodicule: Minute scale within a grass flower. Thought to represent vestigial petals.

margins: (of a leaf) the edges of the leaf blade; may be lobed, toothed, or entire.

monocot: A group of flowering plants; characterized by parallel leaf veins, flowers with parts in multiple of three, e.g. grasses, orchids, daffodils, palms.

node: The part of a plant stem from which one or more leaves emerge, often forming a slight swelling.

ocrea: A tubular sheath formed from the fusion of two stipules, as in the buckwheat family.

ovary: The hollow base of the carpel of a flower, containing one or more ovules.

ovule: The part of the ovary that contains the female germ cell and, after fertilization, becomes the seed.

palmate: (of a compound leaf) Having five or more leaflets that radiate from one point, handlike.

panicle: A much-branched inflorescence.

petals: Parts of the corolla of a flower, enclosing the stamens and carpels, often colorful.

petiole: the stalk that attaches the leaf blade to the stem.

pinnate: (of a compound leaf) Having leaflets arranged along a central stalk, typically in pairs opposite each other and feather-like.

pistil: The female organs of a flower; same as carpel, but can also denote several carpels fused together into one unit.

raceme: Inflorescence with separate flowers, attached by short stalks along a central stem. The flowers at the base of the central stem develop first. See also *spike*.

receptacle: The tip of the stem that bears the floral parts, sepals, petals, etc.

rhizome: A rootlike stem that grows horizontally under the soil, or on the surface.

samara: A winged nut or achene containing one seed, as in ash and maple.

schizocarp: A dry fruit that splits into single-seeded parts when ripe.

sepals: Parts of the calyx of a flower, enclosing the petals and typically green and leaflike.

simple: (of a leaf) Single, undivided leaf. See also *compound*.

spadix: The fingerlike central part of an inflorescence of members of the arum family, bearing numerous flowers at its base. See also *spathe*.

spathe: A bract-like structure surrounding the inflorescence of members of the arum family. See also *spadix*.

spike: Inflorescence with separate flowers attached directly to the central stem.

stalks: The flowers at the base of the central stem that develop first. See also *raceme*.

stamen: The male fertilizing organ of a flower, typically consisting of a pollen-containing anther and a filament.

staminode: Infertile stamen, sometimes secreting nectar, as in the witch hazel family, or transformed into a petal-like structure, as in the ginger family

stigma: The tip of a carpel that receives pollen during pollination.

stipule: A small leaflike appendage, typically borne in pairs at the base of the leaf stalk.

stolon: A horizontal stem that grows along the soil surface, producing new plants at the nodes, as in grasses and strawberries.

tepals: Sepals and petals that look the same. Common in lilies and magnolias.

tuber: An underground storage unit formed from stems, as in potatoes, or roots, as in dahlias.

umbel: Inflorescence with separate flowers all attached at the same point, umbrella-like.

Index

Bibliography

Books

Beentje, H. *The Kew plant glossary: an illustrated dictionary of plant terms.* Second edition. Kew Publishing, 2016

Byng, J.W. *The Flowering Plants Handbook: a practical guide to families and genera of the world.* Plant Gateway, 2014

Friis, E.M., Crane, P.R. & Raunsgaard Pedersen, K. *Early Flowers and Angiosperm Evolution.* Cambridge University Press, 2011

Heywood, V.H. (consultant editor). *Flowering Plants of the World.* BT Batsford Ltd, 1993

Heywood, V.H., Brummitt, R.K., Culham, A. & Seberg, O. *Flowering plant families of the world.* Kew Publishing, 2007

Hickey, M. & King, C. *The Cambridge Illustrated Glossary of Botanical Terms.* Cambridge University Press, 2004

Judd, W.S., Campbell, C.S., Kellogg, E.A. & Stevens, P.F. *Plant systematics: A phylogenetic approach.* Sinauer Associates, Inc., 1999

Mabberley, D.J. *Mabberley's Plant-Book.* Cambridge University Press, 2008

Websites:

Angiosperm Phylogeny Website, version 13
http://www.mobot.org/MOBOT/research/APweb/

RHS Horticultural database
http://apps.rhs.org.uk/horticulturaldatabase/

RHS Images Collection
http://www.rhsimages.co.uk/

Image credits